Polymer Degradation

Polymer Degradation

by

Tibor Kelen

VNR VAN NOSTRAND REINHOLD COMPANY
NEW YORK CINCINNATI TORONTO LONDON MELBOURNE

Library of Congress Catalog Card Number: 82-4426
ISBN: 0-442-24837-7

Published by Van Nostrand Reinhold Company Inc.
135 West 50th Street, New York, N.Y. 10020

Van Nostrand Reinhold Publishing
1410 Birchmount Road
Scarborough, Ontario M1P 2E7, Canada

Van Nostrand Reinhold Australia Pty. Ltd.
17 Queen Street
Mitcham, Victoria 3132, Australia

Van Nostrand Reinhold Company Limited
Molly Millars Lane
Wokingham, Berkshire, England

15 14 13 12 11 10 9 8 7 6 5 4 3 2 1

Library of Congress Cataloging in Publication Data

Kelen, Tibor.
  Polymer degradation.

  Includes bibliographical references and index.
  1. Polymers and polymerization–Deterioration.
I. Title.
TA455.P58K44 1982   620.1'920422   82-4426
ISBN 0-442-24837-7   AACR2

Foreword

The stability of polymers has always been an important consideration in their applications and, therefore, has been the subject of extensive testing and research. In this book Dr. T. Kelen has incorporated a wide range of topics dealing with polymer degradation as well as stabilization for conditions most likely to be experienced in practical use. The material was assembled initially as course notes for a special lecture series presented by Dr. Kelen during his stay as a visiting professor in polymer science at The University of Akron. Because of his extensive research experience and interest in the subject, he was encouraged to extend the topics covered and to organize the material for publication, which he has done admirably. We expect that Dr. Kelen's efforts in this respect will provide a valuable reference for students and researchers as well as for those concerned with environmental stability of polymers in a variety of applications.

FRANK N. KELLEY

Professor and Director
Institute of Polymer Science
The University of Akron

Preface

Polymer degradation is one of the most important areas of polymer chemistry, in both scientific and technological respects. This fact is reflected by the increasing number of publications and conferences in the field. Recently, specialized periodicals such as *Polymer Degradation and Stability* and series of books such as *Development in Polymer Degradation* (edited by N. Grassie) and *Developments in Polymer Stabilization* (edited by G. Scott), have appeared which try to meet the growing demand for knowledge collected about the various types of aging, degradation, and stabilization processes. However, these journals and books do not give a systematic treatment of the whole field, but rather contain reports or reviews on specific topics of polymer degradation according to the interest and activity of the individual authors. The situation is the same for textbooks and monographs. *Aspects of Degradation and Stabilization of Polymers* (edited by H. H. G. Jellinek) includes 14 chapters written by 15 authors on selected topics. Another approach represents the other extreme, namely, the very circumstantial description of an exclusive degradation type such as *Photodegradation, Photo-oxidation and Photostabilization of Polymers* by R. Rånby and J. F. Rabek.

The primary purpose of the present book is to provide graduate students with an introduction to polymer degradation and to help interested workers in scientific research and industry to solve related problems. The book will also be useful in courses on polymer chemistry. This book treats polymer degradation quantitatively whenever possible but without going into excessive theoretical detail. Emphasis is placed on the structural and physicochemical factors which affect and control degradation processes, and where practicable, a kinetic approach is applied to describe degradation events. In order to familiarize the reader with practical aspects of degradation studies, methods are described which may help them to commence such investigations.

Although the most important types of polymer degradation are dealt with in detail, some others are described very briefly or are not even included in the book. For example, problems connected with the burning of polymers are mentioned only in respect to stabilization. Also not included, or only touched upon, are the degradation problems of natural and synthetic rubbers and thermoset polymers. Selections have been made on the basis of the relative importance of the specific degradation type or polymer, or because the polymer in question (e.g., rubber) has its own abundant, rather separate, degradation literature.

Specific references to the material are not given; however, as in most textbooks, reading material is listed at the end of each chapter. These lists are not complete surveys of the literature of polymer degradation, they were compiled by selecting those books and articles which were actually used in the formulation of facts and theories included in the book.

The book originated from the author's lectures on polymer degradation at the Eötvös Lóránd University, Budapest, Hungary and at The University of Akron, Akron, Ohio. The author has used, whenever possible, the results of his own research on PVC and polyolefin degradation, and is very grateful to the Central Research Institute of Chemistry (CRIC), Hungarian Academy of Sciences, Budapest, Hungary for providing for many years conditions amenable to the execution of his work. It is a pleasure for the author to thank all of his colleagues in CRIC who took part in this research, especially Professor F. Tüdős, as well as Drs. G. Bálint (photostabilization), M. Iring (polyolefin oxidation), T. T. Nagy (PVC degradation), and others.

The author would like to express his gratitude to the Institute of Polymer Science (IPS), Akron, Ohio for giving him the opportunity to write the book. He is especially grateful to Professor F. N. Kelley, director of the IPS, for his continual encouragement, advice, and help. The author is greatly indebted to Ms. M. Israel and L. Hutchinson for typing the manuscript. Special thanks to Mr. W. M. Ferry for his help in correcting the author's poor English; of course the author accepts responsibility for any linguistic errors present in the final text.

TIBOR KELEN

Contents

Foreword (*Frank N. Kelley*)   v
Preface   vii

1. General Aspects of Polymer Degradation   **1**

    1.1 Polymer Life Phases and Degradation   2
    1.2 Factors and Stresses Causing Degradation   3
    1.3 Factors Affecting Polymer Stability   4
    1.4 Types of Polymer Degradation   8
    References   9

2. Methods of Studying Polymer Degradation   **10**

    2.1 Methods for Testing the Weathering of Polymers   11
        Outdoor Methods   11
        Laboratory Methods of Weathering   13
    2.2 Iso- and Anisothermal Methods of Accelerated Aging   13
    2.3 Methods of Investigating Thermo- and Photooxidative
        Degradation   21
    2.4 Product Analysis Methods   35
    2.5 Extrapolation and Prediction of Service Life   40
    References   42

3. Depolymerization   **43**

    3.1 Chemical and Thermodynamic Aspects of Depolymerization   43
    3.2 Kinetics of Depolymerization   50
    3.3 Ceiling Temperature   54
    References   57

4. Random Chain Scission and Cross-linking 58

4.1 General Features 58
4.2 Kinetic Treatment of Chain Scission 60
4.3 Probability Treatment of Chain Scission 62
4.4 Random Cross-linking; Simultaneous Chain Scission and
Cross-linking 68
References 71

5. Degradation Without Chain Scission 73

5.1 General Remarks 74
5.2 Thermal Degradation of PVC 77
Primary Processes 78
Secondary Processes 85
5.3 Thermo-oxidative Degradation of PVC 96
5.4 Degradation of PVC under Dynamic Conditions 102
References 105

6. Oxidative Degradation 107

6.1 General Remarks 108
6.2 Effect of Polymer Properties on Oxidation 114
6.3 Kinetics of Polyolefin Thermal Oxidation 116
6.4 Some Differences in the Oxidation of Polyethylene
and Polypropylene 131
References 135

7. Photodegradation 137

7.1 Photophysical Processes 137
7.2 Photochemical Processes 141
7.3 Photooxidation 143
7.4 Sensitized Photodegradation 147
References 150

8. Biodegradation 152

8.1 Methods of Testing Biodegradability 153
8.2 Biodegradability of Polymers 154

8.3 Biodegradation of Commonly Used Additives   155
8.4 Biodegradable Polymers   155
References   156

9. **Mechanical Degradation**   **157**

9.1 Methods of Investigation   157
9.2 Molecular Mechanism and Theories   163
9.3 Factors Affecting Mechanical Degradation   168
References   171

10. **Stabilization of Polymers**   **173**

10.1 General Requirements   174
10.2 Antioxidants   175
        Chain Breaking Antioxidants   175
        Preventive Antioxidants   182
10.3 Photostabilization   185
        Light Screens   186
        Ultraviolet Absorbers   187
        Quenchers   188
10.4 Heat Stabilization of PVC   193
        Calcium/Zinc Stabilizers   194
        Barium/Cadmium Stabilizers   196
        Lead Stabilizers   197
        Organotin Compounds   198
10.5 Other Applications   199
        Stabilization Against Burning—Flame Retardants 199
        Antiozonants   200
        Stabilization Against Biodegradation   200
References   201

Index   **203**

Chapter 1
General Aspects of Polymer Degradation

What is "polymer degradation"? It is the collective name given to various processes which degrade polymers, i.e., deteriorate their properties or ruin their outward appearance.

Generally speaking, polymer degradation is a harmful process which is to be avoided or prevented. The operation which can be undertaken to inhibit or to retard degradation is called polymer stabilization. In order to do this suitably, with maximum efficiency, we must understand the mechanism of polymer degradation, we must identify the factors and stresses causing it, and we must know what factors affect polymer stability.

Sometimes, although not very often, polymer degradation may be useful. Depolymerization leading to high purity monomers may be exploited for practical production of such materials. Another important field in which degradation is desirable is the destruction of polymeric waste materials.

Within the scope of this book we will deal with both aspects mentioned. Of primary importance, of course, are the problems connected with the damaging character of degradative processes. It is not possible to include all details in this book. From the great variety of polymer degradation processes only the most important types will be treated here, demonstrated in connection with a few selected industrial polymers, e.g., polyvinyl chloride and polyolefins. The reason for the selection is the personal experience of the author which has resulted from his own research.

1.1 POLYMER LIFE PHASES AND DEGRADATION

Degradation may happen during every phase of a polymer's life, i.e., during its synthesis, processing, and use. As mentioned before, even after the polymer has fulfilled its intended purpose, its degradation may still be an important problem.

Depending on polymerization conditions, during polymer *synthesis* depolymerization may take place. Depolymerization is the inverse of polymerization, namely, a stepwise separation of the monomers from the growing chain end. As we will see later, polymerization is possible only below the equilibrium temperature of the system, the so-called ceiling temperature. From a practical point of view, it is important that the eventual fate of a polymeric material is often decided during its synthesis. Depending on the polymerization technology, varying amounts of "weak sites" may be built into the polymer which will later occasion its deterioration. For example, the presence of tertiary chlorines in PVC resulting from branch formation during polymerization may reduce PVC stability. During synthesis, various contaminants such as catalyzer residues or other polymerization additives may enter the polymer. The presence of impurities may be very decisive with respect to polymer life span. The amount of additives used during PVC polymerization techniques increases in the sequence block–suspension–emulsion; the stability of the polymer decreases in the same sequence.

During *processing*, the material is subjected to very high thermal and mechanical stress. These drastic stresses may initiate a variety of polymer degradation processes leading to a deterioration of properties even during processing. On the other hand, the damaging of the material may result in the introduction of various defects in the polymer which will work as degradation sources during its subsequent service life. For example, in the presence of oxygen traces during processing, carbonyl groups can be formed in polyolefins; these will later absorb UV light during outdoor applications and thus will function as built-in sensitizers for photodegradation processes. In PVC, a small amount of HCl may be eliminated from the polymer during processing. This alone does not influence the properties very much, but the resultant double bonds and the allyl-activated chlorines joined with them are very dangerous sources of PVC degradation.

Because of the dual nature of degradation during processing, i.e., deterioration and introduction of defects as potential sources of deterioration, it is very important to add stabilizers to the polymer before processing. This is necessary not only in cases in which processing supplies finished products, but also during intermediate (e.g., granule) production.

The best known appearance of polymer degradation is connected with the *use* of these materials. Some kinds of polymer degradation, such as the outdoor aging of PVC roofs or the stiffening and discoloration of badly composed vinyl handbags, etc., are well known. The scale of polymer *applications* is, however, very broad; consequently, the stability requirements are highly diverse. In most cases there is a demand for a long service life; sometimes only a predicted (usually short) lifetime is required. Some types of mulch foils for horticultural applications may be completely destroyed after their useful life of several weeks or months. Underground cables for telecommunication or heavy current applications are, however, expected to last several decades. In both of these examples, the polymers are in contact with the soil and the same types of bacteria probably attack them; the biodegradability of the polymers employed should thus be very different.

The problem of *waste disposal* is increasing with the use of increasing amounts of plastic materials. Organized recovery presently exists only in the case of production wastes inside polymer factories. In domestic garbage there is an every-increasing portion of plastic wastes, the destruction of which requires expensive equipment. During the burning of one pound of PVC, approximately 160 liters of HCl are evolved, which is very undesirable because of air pollution. Protection of the environment requires improved packaging materials which are "self-destructing," i.e., which will be degraded very rapidly when exposed to the effects of sunlight, humidity, and — finally — soil bacteria.

In summary, we can conclude that degradation plays an important role in every life phase of a polymer. Thus the importance of studying its mechanism, relationships, etc., is quite obvious.

1.2 FACTORS AND STRESSES CAUSING DEGRADATION

Macromolecules are composed of monomeric units which are joined by chemical bonds to each other. The monomeric units contain chemical bonds which either are in the main chain of the macromolecule or connect various atoms or side groups to it. Side groups, if they are present, contain additional chemical bonds. All of these bonds may be reaction sites in polymer degradation, and various energy sources may be effective in supplying the energy necessary to break the bonds. The bond energies are manifold and depend not only on the kind of atoms connected by the bond but also on the chemical and physical characteristics surrounding the bond. The dissociation energies of chemical bonds in common polymers range from about 65 kcal/mol (C–Cl) to 108 kcal/mol (C–F) with carbon-carbon bonds in the middle (75–85 kcal/mol). The most important types of energy that cause polymer degradation are heat, mechanical energy, and radiation. Thermal and mechanical degradation of polymers may occur during thermomechanical processing. An extreme case of heat damage is burning; the flammability and combustion behavior of polymers are very important in many applications. A typical example of pure mechanical destruction of a polymeric material is grinding, although the evolved heat may play a role even in this case. The most common form of radiant energy which causes degradation is that of the UV component of sunlight; the energy of a photon with $\lambda = 300$

nm is about 95 kcal/mol, which is higher than most bond dissociation energies in polymers. In special applications, e.g., polymers used in x-ray laboratories in hospitals or in plastic parts for aerospace applications, nuclear radiation may cause degradation.

Not only is the role of temperature important when the heat necessary for bond dissociation is supplied by thermal motion of the atoms, it can also activate various chemical and biological processes. Reactions of polymers with oxygen and moisture, generally present in various applications, are very important in polymer degradation. Ozone degradation of elastomers is of the utmost importance in the rubber industry. Recently, the stability of polymers in the presence of pollutant gases such as sulfur dioxide and nitrogen dioxide has aroused interest. However, the factors generally are not separated; various combinations of damaging components (e.g., heat, mechanical stress, moisture, and oxygen) may initiate very complex processes. The stresses may be sudden and of short duration (e.g., a hammer blow), but they may also act for a very extended time period. The behavior of the same polymer under impact and fatigue conditions may differ to a great extent; this is also true when other types of stress, not just mechanical, are involved.

As a result of degradation processes, initiated and completed by the above factors or their combination, the internal properties as well as the external appearance of the polymers may change. Chain scission and cross-linking lead to a change of molecular weight distribution; oxidation and other chemical reactions, in the side chains too, cause changes in chemical composition result in discoloration, etc. These primary alterations in the polymer usually cause the deterioration of mechanical and other technically important properties; as a result, the material loses its value and becomes a useless waste. It is, therefore, of great practical importance to know the factors affecting polymer stability.

1.3 FACTORS AFFECTING POLYMER STABILITY

As previously mentioned, the chemical structure of the polymer is of primary importance in respect to its stability. The chemical composition (i.e., what kinds of *chemical bonds* in what sort of arrangement) is in itself a decisive factor. Bond energies between the same atoms are very different depending on the chemical groups to which the atoms belong. A few selected data are included in Table 1.1.

Tertiary and allylic bonds are usually weaker than primary or secondary ones. In polymers consisting only of primary and secondary carbon atoms (e.g., PVC), the presence of such bonds is undesirable because these form weak sites which are very easy to attack. Processes leading to these bonds during polymerization (e.g., PVC branching or dehydrochlorination) are to be avoided.

The dissociation energies of the various bonds in the polymer may determine

Table 1.1. Bond Dissociation Energies of Various Single Bonds.

BOND BROKEN A—B	BOND DISSOCIATION ENERGIES (KCAL/MOL)	BOND BROKEN A—B	BOND DISSOCIATION ENERGIES (KCAL/MOL)
C_2H_5-H	99	$C_6H_5-CH_3$	94
$n\text{-}C_3H_7-H$	98	$C_6H_5CH_2-CH_3$	72
$t\text{-}C_4H_9-H$	91	CH_3-Cl	84
$CH_2=CHCH_2-H$	82	C_2H_5-Cl	81
C_6H_5-H	103	$CH_2=CHCH_2-Cl$	65
$C_6H_5CH_2-H$	83	CH_3-F	108
$C_2H_5-CH_3$	83	C_2H_5-F	106
$n\text{-}C_3H_7-CH_3$	83	$HO-OH$	51
$t\text{-}C_4H_9-CH_3$	81	$t\text{-}C_4H_9O-OH$	36

the course of degradation: the process always begins with the scission of the weakest available bond or with an attack at this site, and the first step usually determines the further direction of the process. Other components of the chemical structure, such as steric factors, stability of the intermediates, or the possibility of their resonance stabilization, may also have great influence on degradation. Such factors may even change the value of the bond dissociation energies.

Table 1.2 shows the effect of steric factors and resonance stabilization on

Table 1.2. Bond Dissociation Energies of Some CH_3-R Bonds.

—R	BOND DISSOCIATION ENERGIES (KCAL/MOL)
$-CH_3$	88.4
$-C_2H_5$	84.5
$-n\text{-}C_3H_7$	84.9
$-n\text{-}C_4H_9$	84.7
$-i\text{-}C_3H_7$	83.8
$-t\text{-}C_4H_9$	80.5
$-CH_2-CH=CH_2$	73.6
$-CH(CH_3)-CH=CH_2$	72.3
$-CH_2-C_6H_5$	71.9
$-CH(CH_3)-C_6H_5$	68.7
$-C(CH_3)_2-C_6H_5$	65.7
$-CH=CH_2$	93.7
$-C_6H_5$	94
$-CH_2-OH$	82.1
$-CH_2-C(O)-CH_3$	79
$-CH_2-CN$	77.1

bond dissociation energies of some CH_3-R bonds. The effect of bond energies on polymer stability is illustrated in Figure 1.1; the temperatures at which the original weight of the polymer samples decreased by 50% after 30 min of heating in a vacuum are shown here as a function of bond dissociation energies. As can be seen, the slope of the line in Figure 1.1 is about 6°C/(kcal/mol), i.e., an increase in the dissociation energy of about 5 kcal/mol causes approximately 30°C increase of the half weight loss temperature.

Some *comonomeric units* incorporated in the copolymers may influence stability. Such units usually modify the application properties of the polymer, for example, the T_g or mechanical strength; they can, however, also improve the stability. Thus, the incorporation of a few percent of dioxolane units into poly-formaldehyde greatly reduces its depolymerization because the dioxolane units inhibit the unzipping of formaldehyde units.

The presence of some comonomeric units, like the presence of certain additives or polymer *blend components* which are not chemically bonded to the polymer (although added in order to improve its properties), may decrease stability. For example, the rubber component of high impact polystyrene usually contains unsaturated bonds which are sources of various degradation reactions. High impact polystyrene products are improved materials from the view-

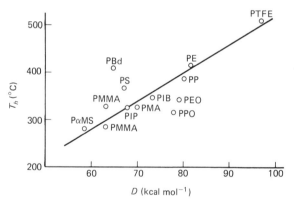

Figure 1.1. Half weight loss temperature (T_h) vs bond dissociation energy (D) for polymers of chain scission type. PTFE, polytetrafluoroethylene (97); PE, polyethylene (81.6); PP, polypropylene (80.0); PEO, polyethylene oxide (79); PPO, polypropylene oxide (78); PIB, polyisobutylene (76.9 – 4* = 73); PMA, polymethyl acrylate (70); PIP, polyisoprene (67.8); PS, polystyrene (66.5); PBd, polybutadiene (64.7); PMMA,† polymethyl methacrylate (65.7 – 4* = 62); PαMS, poly-α-methylstyrene (62 – 5* = 57). The figures in parentheses are the estimated values of D in kcal/mol. (Reprinted, by permission of Elsevier Publishing Company, from reference 3.9.)

*Subtraction due to steric hindrance.
†Two data are given for T_h.

point of their improved mechanical properties, but they are more sensitive to light or oxidation than polystyrene itself. An increasingly utilized method of improving stability is the chemical *modification* of polymers. Elimination of weak sites or substitution of labile groups by stable ones via grafting onto the polymer may be very effective and useful especially when the substituents simultaneously improve other properties. Alkylation of the weak sites in PVC may result in internal plasticization of the material.

The *tacticity* of the polymer plays an important role in the degradation behavior. Atactic and isotactic polypropylene have very different oxidative stability (the isotactic one is much more stable). Syndiotactic PVC prepared at low temperatures has increased stability compared to the ordinary material produced at about 50°C. It is, however, very difficult to separate the effect of tacticity from that of the morphology of the material because a change in tacticity is usually connected with a change of morphology.

Physical and *morphological* factors may also influence polymer stability. It is well known that oxidation is always initiated in the amorphous phase of semi-crystalline polymers and the propagation of the oxidation into the crystalline phase is a result of the destruction of the crystalline order. Thus, crystallinity is an important characteristic of the polymer from the viewpoint of stability.

The morphology of the material is decisive from the point of view of *diffusion* conditions. A compact material is usually more stable against oxidation because the diffusion of oxygen into the product is more difficult than with a material of loose structure. On the other hand, the facile diffusion of HCl evolved from PVC with a loose morphology reduces the autocatalytic character of the degradation which may lead to a catastrophic destruction in the case of compact and dense materials.

Similar to the internal chemical stresses already mentioned (weak sites, etc.), the *internal mechanical stresses* which are left in the material or introduced by finishing operations are very dangerous. Such stresses may serve not only as sources of later mechanical deterioration but also as initiators of, or assistants to, various chemical attacks. This is especially true in cases of stresses with long duration (the so-called stress corrosion of polymers).

The role of *contaminants* is quite obvious and has already been mentioned in connection with the synthesis of polymers. It is obvious that some *additives* intentionally present in the material, such as plasticizers or lubricants, influence the stability of the composite, especially if the oxidizability and biodegradability of such systems are higher than those of the polymer components. Once radicals are formed in the additive, they attack the polymer and vice versa; i.e., composites are sometimes less stable than their components. In special cases, additives are intentionally used to promote degradation of the composites (e.g., photosensitizers or plasticizers) which are specific culture media for bacteria in some rural and horticultural applications. On the other hand, additives may

also stabilize the polymer: the use of antioxidants, photostabilizers, etc., which
we will discuss later in detail, is based on this fact.

1.4 TYPES OF POLYMER DEGRADATION

There are various schemes to classify polymer degradation. Because of its complexity, with regard to both the causes and the response of the polymer, classification is usually performed on the basis of the dominating features. One of the
most frequent classifications has been based on the main *factors* responsible for
degradation:

> thermal, thermo-oxidative, photo, photooxidative, mechanical, hydrolytic,
> chemical, and biological degradation; degradation by high energy radiation;
> pyrolysis and oxidative pyrolysis; etc.

Another possible classification is based on the main processes taking place as
dominating *events* during degradation:

> random chain scission, depolymerization, cross-linking, side group elimination,
> substitution, reactions of side groups among themselves, etc.

Although these basic types of degradation processes seldom take place separately, some understanding may be gained by studying the degree of polymerization vs monomer yield diagrams shown in Figure 1.2. Here we use $P'_n(t)$, the

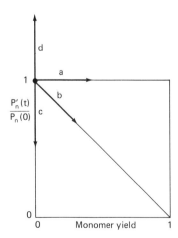

Figure 1.2. Degree of polymerization vs monomer yield diagram of polymer degradation.
(Reprinted, by permission, from reference 1.6.)

number average degree of polymerization of the open system, in which the evolved monomers do not remain in the system and are not taken into account.

Direction a (monomer production without change of P_n) is characteristic for depolymerization with $\nu > P_n$: once unzipping is initiated, the whole macromolecule depolymerizes if the kinetic chain length is higher than the degree of polymerization. Direction b is typical for depolymerization processes with $\nu < P_n$.

Direction c (decrease of P_n without monomer production) corresponds to a random main chain scission process. In the case of direction d, cross-linking is the dominating process. If $P_n'(t)/P_n(0)$ does not change and no monomer is produced, then reactions without chain scission or cross-linking proceed (e.g., side group elimination).

From an experimental point of view such classifications are necessary and useful because they facilitate the study of polymer degradation. The separation of the causes is especially important. Simultaneous application of various damaging factors, although this is usually the case in practice, would make the evaluation of degradation processes extremely difficult.

REFERENCES

1.1. Allara, D. L. and Hawkins, W. L. Stabilization and degradation of polymers. In *Advances in Chemistry Series 169*. Washington: American Chemical Society, 1978.

1.2. Conley, R. T. (ed.). *Thermal Stability of Polymers*. New York: Marcel Dekker, 1969.

1.3. David, C., Geuskens, G. and Rabek, J. F. Degradation of polymers. In (Bamford, C. H. and Tipper, C. F. H., eds.) *Comprehensive Chemical Kinetics*, Vol. 14. Amsterdam: Elsevier, 1975.

1.4. Emanuel, N. M. Structure related kinetic aspects of aging and stabilization of polymers (review). *Polymer Science USSR* 20:2973 (1979).

1.5. Geuskens, G. (ed.). *Degradation and Stabilization of Polymers*. London: Applied Science Publishers, 1975.

1.6. Grassie, N. *Chemistry of High Polymer Degradation Processes*. New York: Interscience, 1956.

1.7. Grassie, N. Degradation. In (Jenkins, A. D., ed.) *Polymer Science*, Ch. 22. Amsterdam: North-Holland, 1972.

1.8. Hawkins, W. L. (ed.). *Polymer Stabilization*. New York: Wiley-Interscience, 1972.

1.9. Jellinek, H. H. G. *Degradation of Vinyl Polymers*. New York: Academic Press, 1955.

1.10. Jellinek, H. H. G. (ed.). *Aspects of Degradation and Stabilization of Polymers*. Amsterdam: Elsevier, 1978.

1.11. Kuzminskii, A. S. (ed.). *The Aging and Stabilization of Polymers*. Amsterdam: Elsevier, 1971.

1.12. Madorski, S. L. *Thermal Degradation of Organic Polymers*. New York: Interscience, 1964.

1.13. Neiman, M. B. (ed.). *Aging and Stabilization of Polymers*. New York: Consultants Bureau, 1965.

1.14. Reich, L. and Stivala, S. S. *Elements of Polymer Degradation*. New York: McGraw-Hill, 1971.

Chapter 2
Methods of Studying
Polymer Degradation

The methods of investigating polymer degradation are as numerous as the processes themselves. In most cases we must use accelerated or intensified testing methods because the natural processes are too slow; unfortunately, this involves the problem of extrapolating the results to real conditions. As a rule, we simulate the factors causing degradation, generally one at a time, and then apply them in an intensified manner. When we study oxidative degradation of polyolefins (a slow process under real conditions, e.g., by storing the material at room temperature in air, with slight natural illumination in which light and dark periods alternate, etc.), we do it at elevated temperatures (110-170°C) *or* with intensive UV light in a pure oxygen atmosphere. To follow the process, i.e., to observe the response of the polymer to the above attacks, we must analyze the system as many ways as possible. Any method which supplies information about the changes in the polymer or its surface or about the intermediates or products remaining in the polymer or evolved from it, may be useful in understanding the polymer degradation mechanism. Of course, investigation of the change of properties due to degradation is important and sometimes informative with respect to process decoding.

It is clear from the above that the methods of studying polymer degradation are strongly dependent on the actual mode of degradation of the specific polymer. In the following material, only a selection of the most common methods will be discussed.

2.1   METHODS FOR TESTING THE WEATHERING OF POLYMERS

An important requirement for polymers is the ability to withstand the adversity of weather. This behavior of polymers is characterized as their "weatherability." The effect of the weather is a result of the joint attack of various constituents, the most important of which are the ultraviolet light of the sun's radiation and the moisture in the atmosphere. Weathering is usually investigated by outdoor exposure or under laboratory conditions. Because of the difference in conditions, it is rather difficult to determine correlations between the results of the two types of experiments.

Outdoor Methods

During outdoor exposure, a diversity of factors act on the polymer. These may occur simultaneously or sequentially. The most important of these is sunshine, the spectral composition and intensity of which are very different in different seasons, at different hours of the day, and at different sites on the earth (depending on both geographical location and elevation above sea level). The chemical composition of the atmosphere, its moisture content, and the various kinds and amounts of pollutants in it are important (and changing) factors. The frequency and intensity of rain and winds, the amount of dust in the wind, the bitterness of winter frosts, etc., significantly influence the extent of weathering, partly by their mechanical (abrasive) action on the material.

Thus, the selection of the exposure site and the mode of sample placement are extremely important with respect to the usefulness of the results. In order to obtain characteristic results representing average conditions and applicable also for estimation of service life, a rather long (3-5 years) exposure is usually necessary. To follow the changes in the course of weathering, samples are taken from the exposed material every 1-2 months (or, in the case of less intensive effects, every 6-12 months). Simple visual observations, measurements of tear strength, and other investigative methods can be applied to estimate the extent of degradation. In order to obtain a reliable characterization in any case, numerous measurements which can be statistically evaluated are necessary. It is advisable to use film samples for outdoor aging because finished products, especially those with complicated forms, may include, to various extents, internal stresses, oriented or crystalline regions, and other anisotropies which result from the preparation process. Thus, weatherability investigations using finished products generally supply poorly reproducible results.

Investigating conditions are prescribed in testing standards such as ASTM D1435; these standards are, however, not uniform enough internationally. Samples are, in general, placed on frames directed to the south, having a 30-45° inclination to the vertical (equatorial mounts). Recently developed equipment uses mirrors to increase the efficiency of outdoor exposure. Such devices are, for example, the EMMA (Equatorial Mount with Mirrors for Acceleration) and EMMAAQA instruments (in the latter the samples are also sprayed with water).

The EMMA equipment (U.S. Patent 2,945,417) follows the movement of the sun by means of mirrors which are movable, turnable, and inclinable in many directions (Figure 2.1). The mirrors are made of aluminum and reflect 70–80% of the ultraviolet radiation. The equipment is automated and very effective. Samples in such instruments receive a radiation dosage about 8–10 times more intense than samples in mounts without mirrors. Therefore, a rather high temperature (about 150°C) develops, and application of air cooling is necessary.

If possible, exposure locations are selected in places where the air is clean and sunshine is frequent, such as (in the United States) Florida and Arizona. Weatherability results obtained at different locations are difficult to compare, but the data can be evaluated if the exposure conditions are specified exactly. In general, the location, the duration of exposure, and the average temperature

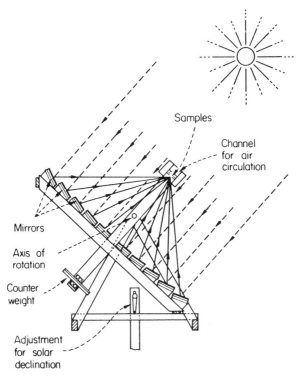

Figure 2.1. Schematic representation of the EMMA device. [From Caryl, C. R.: U.S. Patent 2,945,417 (1960).]

and humidity are the most important characteristics of an outdoor weathering test.

Laboratory Methods of Weathering

Weather-Ometers are laboratory devices for *accelerated* testing of the weatherability of polymers. These instruments produce, in an intensified manner, the climatic effects occurring in nature. The most important constituents, i.e., UV radiation and humidity, are simulated. In a typical Weather-Ometer, the polymer samples are mounted on a rotating cylinder. The light source, usually a carbon-arc or xenon lamp, is located at the center of the cylinder. In nature, day and night alternate; this is simulated by means of a fixed baffle which, by choice of proper size, provides light and dark periods of sufficient duration. From time to time, water is sprayed on the samples. In modern instruments, spray intervals, temperature (35–75°C), and radiation intensity are automatically regulated and recorded. An American version of the Weather-Ometer is the Atlas Type XI apparatus. Other widely used equipment includes the Original Hanau (West Germany) Xenotest Type 150, 450, and 1200 Weather-Ometers. The light source in these devices is a xenon lamp, whose radiation is an excellent approximation of the spectral composition of sunlight.

The physicochemical and mechanical measurements which constitute the ground for evaluation of polymer degradation in the case of both outdoor and laboratory weathering are not specific and will be discussed later.

2.2 ISO- AND ANISOTHERMAL METHODS OF ACCELERATED AGING

Thermal methods are widely used in the assessment of polymer stability. The availability of commercial measuring devices and the ease with which large quantities of data may be accumulated have, unfortunately, led to an increase in the volume, rather than the quality, of this kind of research on polymer degradation reactions. There is also an abuse of the term "stability" in this field: statements such as "stable to 500°C" or "stability was assessed by thermogravimetry"—with little or no indication of the experimental conditions—are quite frequent. Often there is no reference given in reports to the transitions that the material may have undergone prior to weight loss and there is not characterization of the material given other than that it was stable in terms of weight loss at the treatment temperature. On the other hand, it sometimes actually is not possible to derive sufficient information from thermal studies without combining them with other investigative methods.

Isothermal weight loss measurement was a traditional experimental tool in the pioneering work on polymer degradation. As a matter of fact, to study the kinetics and mechanism of degradation processes, isothermal conditions are the best. Our interest is usually not restricted to the weight loss or increase. Sometimes, such changes do not occur at all. Often, the change of temperature not only causes the reaction rates to change but also alters the mechanism. Processes which have no observable effect at low temperatures may become dominant with increasing temperature. Information on activation energies may be obtained from isothermal measurements at various temperatures. The evaluation of the kinetics of degradation processes is much easier and more straightforward with isothermal measurement than with other methods of thermal analysis.

Thermal analysis may be defined simply as the measurement of a property of a material as a function of temperature. The most important anisothermal methods which are relevant to polymer stability studies include thermogravimetry (TG), differential thermal analysis (DTA), its modern version, differential scanning calorimetry (DSC), and thermal volatilization analysis (TVA).

Thermogravimetry is the measurement of the change in weight of a substance as the temperature of its environment is varied in a controlled manner. The essential variables measured in a TG experiment are weight, time, and temperature. With a modern thermobalance, the measurement of weight need not be a precision-limiting factor. The precise measurement of time has never been an experimental problem. However, temperature measurement has remained a serious problem, and careful attention must be given to it in thermal analysis systems. A temperature sensor should be in contact with the degrading specimen whenever possible. Temperature calibration may be substituted for direct contact, but changes in factors such as the pressure in the environment of the sample or the endo- or exothermic character of the reaction may invalidate such a calibration.

There are many TG systems commercially available; as an example, we will discuss here the Mettler thermoanalyzer in which the balance operates on the substitution principle with built-in counterweights and electromagnetic compensation (Figure 2.2). An extremely important advantage of this type of apparatus is that the thermocouple is in contact with the sample during the weight change measurement. An additional feature of the system is versatile atmospheric control. It may be used at pressures as low as 4×10^{-5} torr (5×10^{-3} pascal) or at atmospheric pressure with gaseous flow rates of 3 ml/sec. A special reactive gas inlet allows rapid changes of atmosphere in the vicinity of the sample, thus permitting the application of corrosive gases without harming the balance mechanism. The thermo-analyzer usually simultaneously supplies the derivative of the TG curve (DTG) which immediately shows the rates of the weight changes.

Differential thermal analysis is the measurement of the difference in temperature between a substance and a reference material as the temperature of their environment is varied in a controlled manner. If an exothermic process takes

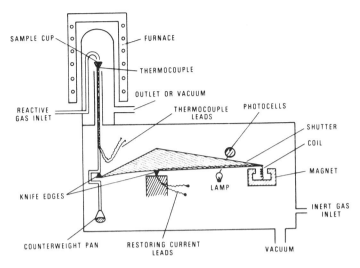

SAMPLE CUP

FURNACE

THERMOCOUPLE

OUTLET OR VACUUM

REACTIVE
GAS INLET

THERMOCOUPLE
LEADS

PHOTOCELLS

SHUTTER

COIL

MAGNET

LAMP

KNIFE EDGES

INERT GAS
INLET

COUNTERWEIGHT PAN

RESTORING CURRENT
LEADS

VACUUM

Figure 2.2. Substitution balance (Mettler thermoanalyzer) for TG measurements. (Reprinted, by permission of Elsevier Publishing Company, from reference 1.10.)

place in the sample material, its temperature will increase with respect to the temperature of the reference. An endothermic process in the sample will have the opposite effect on the temperature difference. As shown in Figure 2.2, systems have usually been adapted to simultaneous DTA measurement and will simultaneously display TG, DTG, and DTA curves, as well as pressure, temperature, and gas velocity. An additional sample cup with the reference material is located very close to the polymer sample; the enthalpically stable reference material should be inert to the applied atmosphere and must not undergo any endo- or exothermal changes in the temperature range studied.

A schematic DTA curve illustrating many of the typical features is shown in Figure 2.3. The temperature difference between the sample and reference, ΔT, is plotted as a function of time or temperature at a constant rate of heating. A shift in the baseline may result from a change in the heat capacity (or mass) of the sample. The peak area is proportional to the enthalpy change in the material; the peaks in Figure 2.3 are, of course, idealized.

Much information about the degradation process can be obtained from analysis of DTA curves. An advantage of DTA is that reactions occurring without weight loss may also be detected. Often, such reactions are responsible for the deterioration of material properties. However, because of the uncertainty in interpreting DTA thermograms, there has been only limited application of DTA to aging studies of polymers.

Differential scanning calorimetry measures the endo- or exothermic effects in the sample under study but uses a method of heat flow measurement based on

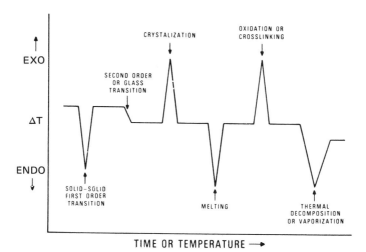

Figure 2.3. Schematic DTA curve. (Reprinted, by permission of Elsevier Publishing Company, from reference 1.10.)

power compensation. A DSC apparatus consists of twin microcalorimeters kept in a stable environment. Each calorimeter contains a temperature sensor, a heater, and a container for the sample or reference material. The sample and reference calorimeters are maintained at approximately the same programmed temperature by addition of electric power to their heaters. The difference in power supplied to the two calorimeters measures the rate of energy change in the sample and is recorded as a function of time.

It should be mentioned that the equipment used in thermal analysis may be employed for purposes other than the study of polymer degradation. DSC devices are widely used in determining phase transitions, T_g values, etc. An advantage of DSC instruments in the investigation of polymer reactions is that the sensitivity of the calorimetric response is not dependent upon heating rate. Investigations can be made at very slow rates of temperature change for quite small samples without loss of sensitivity.

Sometimes, curves from DTA and TG are quite similar. This is true if both techniques are used to measure a system in which the reaction coordinate is a simple function of the weight change of the sample and no other reactions occur. If one considers that the tests measure two fundamentally different properties, it is not unexpected that in the case of complex reactions quite different curves may be obtained by the two techniques. Because DTA can detect reactions in the condensed phase, DTA curves often are more detailed than TG curves.

Often it is found that when the same process is studied by DTA and TG, the results are not comparable. The discrepancy between TG and DTA results can be attributed to the presence of competitive reactions; sometimes the discrepancy may even help to unravel the degradation mechanism.

In order to illustrate evaluative methods of thermal analysis, we will develop here a brief summary of *nonisothermal kinetic analysis*. It is convenient to define the conversion, α, as the reaction coordinate; for a weight loss system this dimensionless quantity is given by

$$\alpha = \frac{W_0 - W}{W_0 - W_f} \tag{2.1}$$

where W stands for the actual, W_0 for the initial, and W_f for the final weight of sample. The relative rate of weight loss is then

$$-\frac{1}{W_0 - W_f} \cdot \frac{dW}{dt} = \frac{d\alpha}{dt} = \frac{d\alpha}{dT} \cdot \frac{dT}{dt} = \beta \frac{d\alpha}{dT} \tag{2.2}$$

where β stands for the rate of temperature change, usually kept constant during a thermal analysis experiment. The simplest and most frequently used model in the analysis of TG data is that in which a proportionality between the reaction rate and the amount of the material still present is assumed:

$$\frac{d\alpha}{dt} = k(1 - \alpha)^n \tag{2.3}$$

where n, in analogy to homogeneous chemical kinetics, is called the order of reaction, and k is the rate constant. The classical model for the temperature dependence of the latter is usually given by the Arrhenius equation; thus with Eqs. (2.2) and (2.3) we have:

$$\frac{d\alpha}{dT} = \frac{A}{\beta} \exp\left(-\frac{E}{RT}\right)(1 - \alpha)^n \tag{2.4}$$

which can be integrated or differentiated to give equations describing the actual experimental curve.

For example, using the logarithm of Eq. (2.4), Freeman and Carroll describe TG data by a difference equation:

$$\Delta \ln\left(\frac{d\alpha}{dT}\right) = -\frac{E}{R}\Delta\left(\frac{1}{T}\right) + n \, \Delta \ln(1 - \alpha) \tag{2.5}$$

for cases when A and β are constant, i.e., not varying with temperature. A plot of $\Delta \ln(d\alpha/dT)/\Delta(1/T)$ vs $\Delta \ln(1 - \alpha)/\Delta(1/T)$ yields n and E/R as slope and inter-

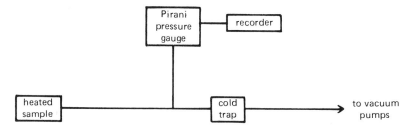

Figure 2.4. Basic layout for thermal volatilization analysis. (Reprinted, by permission of Applied Science Publishers Ltd., from reference 2.12.)

cept. This has become the most popular method for the analysis of anisothermal data. However, great care should be exercised in attaching significance to n and E in cases to which Eq. (2.3) does not apply. Since the simple power law description is usually only an approximation, the Freeman-Carroll method has, like any other which is based on Eq. (2.3), quite limited applicability to polymer degradation kinetics.

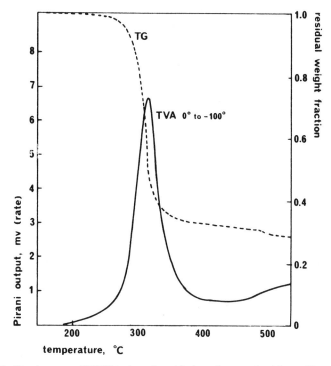

Figure 2.5. Simultaneous TG/TVA data for chlorinated natural rubber. (Reprinted, by permission of Applied Science Publishers Ltd., from reference 2.12.)

Thermal volatilization analysis imposes a programmed temperature increase on the sample and measures the increase of the system's pressure caused by the presence of volatile degradation products. Even in a well-evacuated system, a small pressure buildup will be caused by the passage of volatile products from the heated sample to the cold trap surfaces (Figure 2.4). For small amounts of polymers, the pressure developed is quite low ($\sim 10^{-4}$ -10^{-1} torr), but is measureable using a Pirani gauge.

The change of pressure in a TVA experiment depends on the rate of sample volatilization; hence, a TVA curve resembles a derivative thermogravimetric (DTG) curve. There are, however, several important differences in DTG and TVA curves. While DTG measures the rate of loss due to volatilization of any material in the sample, TVA is sensitive only to those products which are volatile under vacuum at ambient temperature: the less volatile products would condense when leaving the heated area of the apparatus long before they ever reach the Pirani gauge.

Figure 2.5 shows a TG/TVA trace for degradation of a chlorinated natural rubber (64.5% Cl). The degradation behavior of this polymer is similar to that of PVC. Thus, the TVA curve (which is due solely to hydrogen chloride below 400°C) greatly resembles the derivative of the TG curve.

Because of the wide use of thermal methods in stability assessment, various *empirical stability parameters* have been derived from these techniques. A few examples are:

1. From weight loss experiments made under *isothermal* conditions at a specific temperature:
   Induction period of the onset of weight change (t_i)
   Weight loss in some specific time (e.g., in 30 min, W_{30})
   Time for 50% decomposition ($t_{50\%}$)
2. From *anisothermal* weight loss experiments:
   Temperature for 50% decomposition ($T_{50\%}$)
   Residue (%) at a specific temperature
3. From *thermal volatilization* analysis:
   Onset temperature for production of volatiles (T_i)
   Temperature of the maximum rate of volatile production (T_{max})
4. From *DTA* or *DSC* measurements:
   Transition temperatures (T_g, T_c, T_m)
   Incipient heat absorption or evaluation temperature

In addition to these parameters, the activation energy (E) and the preexponential factor (A) may also be regarded as empirical stability parameters because they usually characterize the overall process rather than the elementary reactions.

The study of the high temperature degradation – *pyrolysis* – of polymers is very important. Because of the decisive role of volatile pyrolysis products in combustion processes, in flame spread, and in gas phase reactions accompanying burning, complex instrumentation is usually applied to such studies. Examples include gas chromatographic and mass spectrometric devices. We will deal with these techniques in connection with the methods of product analysis.

A group of thermoanalytical techniques based on the change in mechanical properties of the material is referred to as thermomechanometry.

One very common technique used in polymer science, called *thermomechanical analysis* (TMA), applies a static load. Other methods used to measure mechanical properties of polymers as a function of temperature are carried out using dynamic loading. Both techniques are suitable for evaluating polymer degradation because the change of thermomechanical curves after aging is a measure of the stability of the polymer.

TMA measures the deformation, or actually the deflection in one dimension, of the material under a static load at a uniform rate of heating. The deformation (or deflection) vs temperature curve shows the overall thermal behavior of the polymer in the temperature range studied. The deformation changes rapidly at T_g, and thus TMA (or the differential method, called DTMA) is often used to determine the transition temperatures. TMA is also employed to measure expansion coefficients and thermal dimensional stability, and to detect processing and thermal history effects. There are many parameters which influence TMA of polymers. Some important ones are:

1. TMA variables: probe type, heating/cooling rates, load, sample history, etc.
2. Polymer characteristics: composition, MW/MWD, cross-linking, plasticizers/diluents, microstructure, etc.

The usual TMA probe configurations are expansion, penetration, tension, and the dilatometer mode.

A typical TMA (penetrometer) curve of an elastomeric material is shown in Figure 2.6. The TMA curve resembles the stress relaxation master curve composed from isothermal measurements via the time-temperature superposition principle.

The traditional measurements of dynamic mechanical properties, for example with a torsional pendulum, are widely applied for evaluating thermal properties of materials. A definite shape of the sample is required for such measurements, but it is difficult to keep such a shape constant during aging. An interesting technique has been developed in which the polymer sample is impregnated into interstices of a glass fiber braid. The torsional oscillation of the composite braid gives the relative rigidity modulus and the damping factor of the polymer over a wide range of temperatures. By degradation of the polymer sample, the

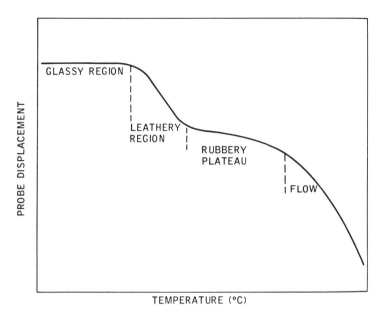

Figure 2.6. Typical TMA (penetrometer) curve of an elastomer. (Copyright 1978 The Franklin Institute Press, from *Thermal Methods in Polymer Analysis*, edited by S. W. Shalaby. Reprinted by permission of the Publisher.)

relative rigidity and damping index vs temperature curves change significantly; this method is called torsional braid analysis (TBA).

2.3 METHODS OF INVESTIGATING THERMO AND PHOTOOXIDATIVE DEGRADATION

During oxidation of polymers various reactions take place between the polymer and gaseous oxygen. The reactions consume oxygen and lead to the absorption of oxygen from the surroundings by the polymer. Thus the *measurement of oxygen absorption* is a very important method of investigating any kind of oxidative degradation.

Since the rate of reaction between a polymer and oxygen depends on the concentration of oxygen in the polymer, the oxygen pressure in the system must be held constant during the experiment. This can be realized in two ways: (1) by continuously replacing the consumed oxygen from an oxygen source, e.g., from an electrolytic cell, and (2) by continuously decreasing the reaction volume. In both cases, the decrease of pressure must be detected and an appropriate correction carried out. In order to minimize the changes in the pressure of the system, automatic devices have been developed.

To avoid sealing problems, oxygen absorption apparatuses usually work at atmospheric pressure; the system is first evacuated and then filled with the desired gas or gas mixture, e.g., pure oxygen, air ($\sim$21% O_2), or O_2:N_2 or O_2:Ar mixtures of diverse O_2 partial pressures. It is important that the pressure change be generated only by oxygen absorption. Volatile products of oxidation, such as H_2O, CO, CO_2, etc., should not be present in gaseous form at the site of pressure measurement. This condition can usually be fulfilled by means of various coolers or absorbers; an exception is oxygen itself, which is presumably evolved from bimolecular termination reaction of $RO_2 \cdot$ radicals.

A schematic representation of a manually operated oxygen absorption apparatus used in polyolefin thermal oxidation studies is shown in Figure 2.7. The oxygen uptake of the polymer is determined by volumetric measurements. The volume of the system is decreased with the aid of a burette filled with mercury to the extent necessary to compensate for the pressure decrease. Several layers of thin polymer films are placed on a sample holder in the reaction vessel (9). Water and carbon dioxide formed during degradation are absorbed by BaO, located on a separate holder above the sample in the reaction vessel (9). The equipment is first evacuated to about 5×10^{-4} torr; the reaction vessel (9) and the reference vessel (8), which were previously placed in a cold thermostat (10), are heated under vacuum to the desired temperature. By means of a needle valve (6), both vessels and the burette (7) are filled with oxygen (or a gas mixture) to the required (preferably atmospheric) pressure. From time to time, the pressure decrease caused by oxygen absorption during the reaction is compensated by decreasing the reaction volume. This is realized by raising the Hg level in the burette (7) by means of a micrometer screw.

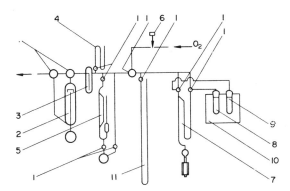

Figure 2.7. Schematic representation of an oxygen absorption apparatus: (1) stopcocks, (2) Hg diffusion pump, (3) trap, (4) shortened manometer, (5) McLeod gauge, (6) needle valve, (7) burette, (8) reference vessel, (9) reaction vessel, (10) thermostat, (11) manometer. (Reprinted, by permission of Pergamon Press, from reference 6.10.)

Many other types of equipment are used for measuring oxygen uptake. An automatic version of volumetric devices using two mercury filled manometers is shown in Figure 2.8. Any decrease in the pressure causes a rise of the mercury level in tubes b and c and a lowering in tube a which breaks an electrical contact in a. A relay starts a servomotor which raises the Hg level in d, thereby also raising the level in a.

A portion of an apparatus for measuring photooxidative oxygen uptake is shown in Figure 2.9. It consists of two irradiation cells C_1 and C_2 coupled to a vacuum, an oxygen supply, and the measuring system as shown earlier. The film sample in C_1 is placed between two brass rings which are screwed together and located on a perforated platform to allow free access of oxygen to the

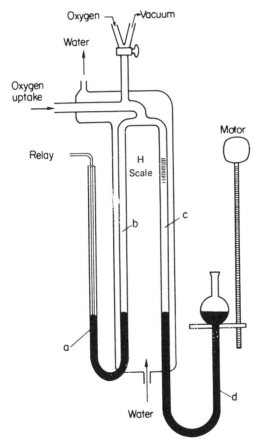

Figure 2.8. Schematic representation of an automatic oxygen uptake instrument. (Reprinted with permission from reference 2.4. Copyright 1968, American Chemical Society.)

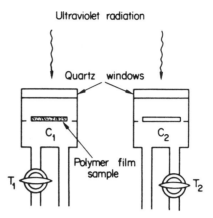

Figure 2.9. Schematic representation of irradiation cells used for oxygen absorption measurements in study of photooxidative degradation of polymers. (From reference 2.6. Copyright 1965. Reprinted by permission of John Wiley & Sons, Ltd.)

underside of the film. Identical equipment (with the exception of the film) is placed in cell C_2. UV light is directed vertically and symmetrically from above.

As already mentioned, some of the products of oxidation are volatile and only a part of them get built into the polymer in the form of carbonyl, carboxyl, hydroxyl, or hydroperoxy groups. Figure 2.10 shows a typical oxygen absorption curve $Z(t)$ for high pressure polyethylene, together with the curve of isothermal weight change $G(t)$. Only at the very beginning of the oxidation do the curves overlap (with common W_0, the initial rate). Later, when the autocatalytic character of oxidation becomes dominant and the decomposition of hydroperoxides leads to chain scission processes, the curves separate: the oxygen absorption keeps to a maximum rate of increase ($W_{Z,max}$), the weight curve to a maximum rate of decrease ($-W_{G,max}$).

The amount of some oxidation products built into the polymer may be determined by means of *infrared (IR) analysis*. This method is widely used for the carbonyl groups which may be present in various forms in the oxidized polymer, e.g., as aldehydes, ketones ($\sim$1720 cm^{-1}), acids ($\sim$1710 cm^{-1}), esters ($\sim$1745 cm^{-1}), and lactones ($\sim$1780 cm^{-1}). The absorption is measured on film samples or on polymer layers supported by pressed KBr discs; in the reference beam, an untreated sample is used. The evaluation of the broad absorption band (1640-1850 cm^{-1}, see Figure 2.11) in terms of specific carbonyl group forms is usually not possible.

The situation is similar in case of OH groups. The broad absorption band in the region 3100-3700 cm^{-1} (see Figure 2.12) is composed of absorptions due to hydroxyl ($-$OH) groups and hydrogen bonded hydroperoxide ($-$OOH)

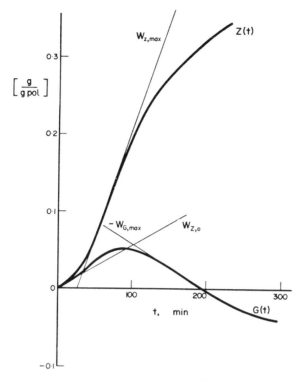

Figure 2.10. Characteristic curves of oxygen absorption, $Z(t)$, and isothermal weight change, $G(t)$, of a high pressure polyethylene sample (approximately 160°C and 760 torr oxygen). (Reprinted, by permission of Pergamon Press, from reference 6.10.)

groups (both absorb at ~3400 cm^{-1}), as well as to free hydroperioxides (~3550 cm^{-1}).

Usually, IR analysis of the samples is only a secondary method for subsequent investigation of oxidized polymers. Sometimes, however, IR absorption spectroscopy is used as the main investigative method of thermo-oxidative degradation. For this purpose a cell has been constructed (Figure 2.13) which allows constant monitoring of the polymer sample undergoing oxidation. The carbonyl and hydroxyl bands shown in Figures 2.11 and 2.12 were measured in such an oxidation cell.

A more direct method of determination of the hydroperoxide content in an oxidized polymer is *iodometry*. The sample is dissolved in a neutral solvent, e.g., 1,2,4-trichlorobenzene, a widely used solvent in oxidation experiments because of its oxidative stability and high boiling point. In hydroperoxide determinations, NaI and acetic acid are added; in dialkyl-type peroxide determina-

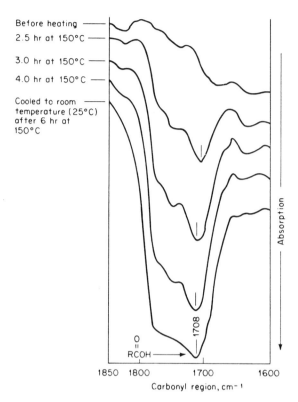

Before heating ——

2.5 hr at 150°C ——

3.0 hr at 150°C ——

4.0 hr at 150°C ——

Cooled to room
temperature (25°C)
after 6 hr at
150°C

Absorption

1708

O
‖
RCOH ⟶

1850 1800   1700   1600

Carbonyl region, cm⁻¹

Figure 2.11. Carbonyl band in a polypropylene sample before and after oxidation (approximately 150°C and 760 torr oxygen). (Reprinted, by permission, from reference 1.14.)

tions, HI is added to the solution. After heat treatment at 110-120°C, the mixture is cooled and is titrated with $Na_2S_2O_3$ solution using starch as indicator. A blank titration is run in parallel with experimental samples.

The carboxyl content of the oxidized polymer may be determined by *acidimetric analysis*. For example, a degraded polypropylene film sample is first refluxed with xylene then titrated with KOH/isopropyl alcohol solution using phenolphthalein indicator. The temperature of the solution throughout titration may not decrease below 100°C. Blank titration must be carried out under identical conditions.

It is known that during certain chemical reactions, e.g., the oxidation of polyolefins, light may be emitted from the system; this phenomenon is called *chemiluminescence*. The luminescence, in general, is associated with the emission of light during electronic transitions from an excited state to the ground state. Besides chemical activation, excitation by high energy radiation is also possible;

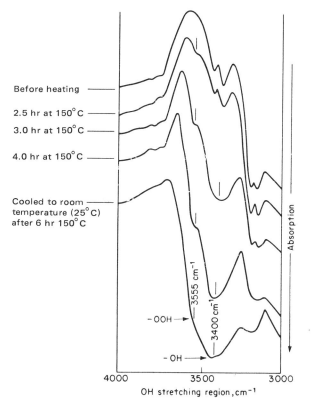

Before heating

2.5 hr at 150°C

3.0 hr at 150°C

4.0 hr at 150°C

Cooled to room
temperature (25°C)
after 6 hr 150°C

3555 cm⁻¹

3400 cm⁻¹

-OOH→

-OH→

Absorption

4000   3500   3000

OH stretching region, cm⁻¹

Figure 2.12. Hydroxyl band in a polypropylene sample before and after oxidation (approximately 150°C and 760 torr oxygen). (Reprinted, by permission, from reference 1.14.)

the luminescence in this latter case is called *fluorescence* (emission by transition from the lowest excited singlet state) or *phosphorescence* (emission by transition from the lowest triplet state).

Figure 2.14 shows the scheme of an apparatus designed for determination of chemiluminescence during thermal oxidation of polymers. The sample can be heated under various atmospheres. The gas flow is regulated by means of needle valves not shown in the figure. The radiation is measured by a photomultiplier tube and a microammeter, and can be recorded. The phototube is thermally isolated from the heated chamber and is cooled by solid CO_2 in order to reduce the noise level and the dark current.

Figure 2.15 shows an arrangement for measuring fluorescence spectra. The incident radiation is generated by a xenon lamp or by a mercury-vapor lamp and passes the excitation monochromator before entering the sample cuvette. The

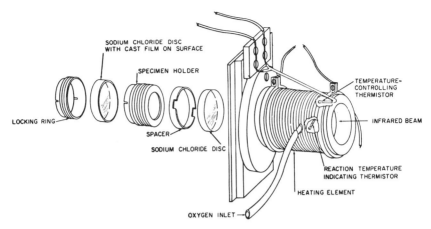

Figure 2.13. Schematic drawing of an oxidation cell used for IR monitoring of degradation. (Reprinted from reference 6.20, p. 22, by courtesy of Marcel Dekker, Inc.)

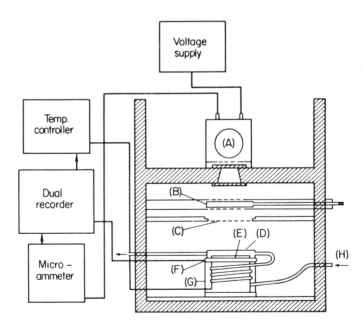

Figure 2.14. Scheme of apparatus used for determination of chemiluminescence: (A) phototube, (B) shutter assembly, (C) filter holder, (D) gas chamber, (E) sample, (F) thermocouple, (G) heating block, (H) gas inlet. (From reference 2.16. Copyright 1964. Reprinted by permission of John Wiley & Sons, Ltd.)

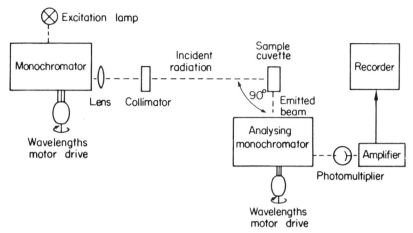

Figure 2.15. Arrangement for recording fluorescence spectra. (From reference 2.14. Copyright 1964. Reprinted by permission of John Wiley & Sons, Ltd.)

emitted beam (i.e., the fluorescence emission) passes the analyzing monochromator and is measured by a photomultiplier tube, the output of which is amplified and recorded. Fluorescence measurements can be carried out in regular spectrophotometers with complementary fluorescence attachment or in specially constructed fluorometers.

Figure 2.16 shows an arrangement for separating phosphorescence from fluorescence. The method is based on the difference in the lifetime of the excited states. The incident and emitted beams are periodically covered by a rotating slotted cylinder. The short-lived fluorescence decays completely during the dark periods, only the phosphorescence radiation comes through. The

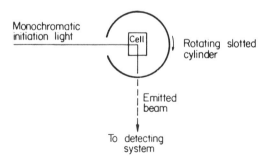

Figure 2.16. Arrangement for separating phosphorescence from fluorescence. (From reference 7.10. Copyright 1975. Reprinted by permission of John Wiley & Sons, Ltd.)

delay can be regulated by changing the rotational speed of the cylinder. The sample is located in a quartz Dewar cuvette maintained at low (77°K) temperature.

Free radicals play an important role in many of the degradation processes. Thermal, photo, or mechanical scission of the polymer chain may produce free radicals. Free radicals are the active centers of oxidation processes, too. Free radicals are compounds containing unpaired electrons; the method of studying their amount and structure is called *electron spin resonance* (ESR) or *electron paramagnetic resonance* (EPR) *spectroscopy*. A magnetic field of sufficient strength causes splitting of the energy levels of the unpaired electrons of the free radical (Zeeman effect), and the magnetic moments are distributed among states with different energies. The energy difference between these so-called Zeeman levels is:

$$\Delta E = g\beta H \qquad (2.6)$$

where g is the spectroscopic splitting factor (for unpaired free electrons, $g = 2.00232$; for organic free radicals, $g = 2.002$-2.004), β is the Bohr magneton (the magnetic moment of the electron), and H is the strength of the applied field.

Transition between the Zeeman levels (Figure 2.17) occurs when energy aborption takes place. This can be achieved by electromagnetic radiation of frequency ν, which satisfies the resonance condition:

$$h\nu = g\beta H \qquad (2.7)$$

where h is Planck's constant. Theoretically, any combination of H magnetic field and ν microwave frequency, satisfying the condition of Eq. (2.7), would be applicable. In practice, a microwave beam of a constant (9–10 GHz) frequency, perpendicular to the main magnetic field, is usually applied and the strength of the magnetic field (about 3200 gauss) is slowly varied. When H reaches the value corresponding to the resonance condition, energy transition occurs [(a) in Figure 2.17] and a band of the electron spin resonance spectrum can be observed. Usually the first derivative (c) of the absorption curve (b) is recorded, but the registration of the second derivative (d) may also be useful.

The number of paramagnetic centers (N_x) in the system is proportional to the absorbed energy, i.e., to the area of the absorption band:

$$N_x \approx \int I(H)\, dH \qquad (2.8)$$

where $I(H)$ is the intensity of microwave radiation.

If the unpaired electrons interact with various nuclei having uniform or dif-

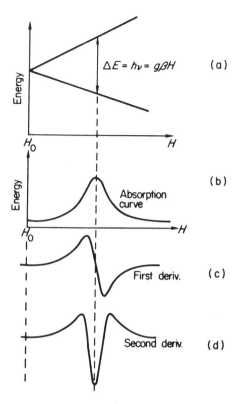

Figure 2.17. Schematic explanation of origin and form of ESR spectra. (From reference 7.10. Copyright 1975. Reprinted by permission of John Wiley & Sons, Ltd.)

ferent magnetic moments, energy absorptions at different H values occur which correspond to the resonance conditions:

$$h\nu_i = g\beta H + \alpha m_i \tag{2.9}$$

where ν_i and m_i are the microwave frequency and the projection of the nuclear spin on the direction of the magnetic field, respectively, corresponding to the i-th transition, and α stands for the hyperfine splitting constant which depends on the measure of electron-nucleus interaction. A few examples of ESR spectra containing several absorption bands because of such interactions are shown in Figure 2.18.

The ESR method has been used in polymer degradation studies to detect and identify free radicals generated by various chain scission processes. The method

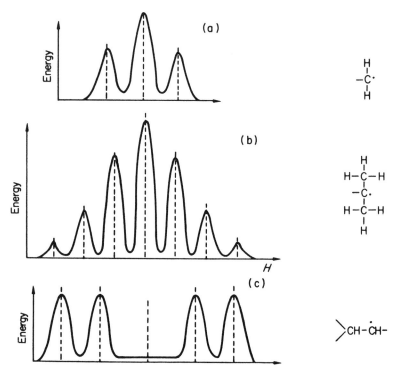

Figure 2.18. ESR absorption spectra for cases in which the unpaired electrons interact with (a) two equivalent protons, (b) six equivalent protons, and (c) two protons of different character. (From reference, 7.10. Copyright 1975. Reprinted by permission of John Wiley & Sons, Ltd.)

can also be applied to determine the number of unpaired electrons, to study the mobility of radicals in the system, and to follow the decay of macroradicals. Commercial ESR spectrometers are available. They usually consist of an electomagnet with variable field strength and a microwave radiation source (klystron). The sample cell is placed in the resonance cavity of the magnet. The apparatus includes regulating, detecting, and recording components as well.

Nuclear magnetic resonance (NMR) spectroscopy is based on similar principles, but with its aid the interactions of the nuclei with their surroundings are studied. The application of NMR is also very important in studying polymer degradation; because of its wide use in polymer chemistry in general we will not deal here with its theory and application.

In studying photochemical reactions the use of various *light sources* and irradiation devices is possible. Irradiation with sunlight (outdoor weathering) and with xenon lamps (in Weather-Ometers) has already been mentioned. Intensive UV radiation can be produced with mercury lamps: low, medium, and high pressure mercury lamps are available, with increasing intensity in this sequence. The two main mercury lines at 1849 and 2537 Å compose these radiations; both have energy high enough to initiate degradation reactions.

Superhigh pressure mercury lamps with forced water cooling are available; these have found wide application in the photochemical industry because of their high radiation power. High emission efficiency is characteristic for the carbon-arc light sources which are widely used as standards for laboratory weathering of paints, dyes, and plastics. Recently, because of the coherence and the extreme monochromaticity of their light, different kinds of lasers have been applied in photochemical investigations. Tunable lasers, covering the entire visible spectrum, are commercially available at the present time.

A few examples of *irradiation devices* have already been mentioned. Figure 2.19 shows a simple apparatus which uses horizontally placed rotating samples. Spherical mirrors are widely used in experimental photochemical techniques; an example of their application is shown in Figure 2.20 in which a broad beam of parallel rays irradiates a large surface. It is to be noted that mirrors for reflecting UV radiation are made from highly polished aluminum, nickel, or silver sheets. Screens coated with magnesium oxide or barium sulfate may reflect about 90% of the incident UV light in the range 2800–4000 Å.

The measurement of the light intensity and the amount (dose) of light absorbed by the sample – i.e., the *dosimetry* – is very important in quantitative photochemistry. In the ultraviolet and visible range, photomultipliers are widely used.

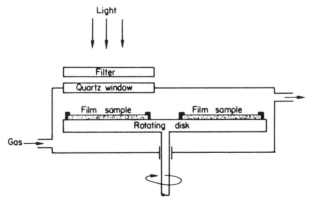

Figure 2.19. Scheme of an irradiation device which employs horizontally placed rotating disc samples in an inert or oxidative atmosphere. (From reference 7.10. Copyright 1975. Reprinted by permission of John Wiley & Sons, Ltd.)

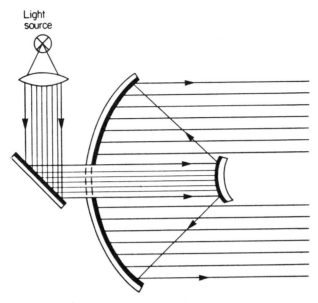

Figure 2.20. Application of mirrors to photochemical experiments. (From reference 7.10. Copyright 1975. Reprinted by permission of John Wiley & Sons, Ltd.)

These operate by application of the photoelectric effect using appropriate scintillators and multiplying the electric current produced by the incident photons; the scintillator-photomultiplier must be calibrated.

The most important parameter describing a photochemical reaction is the quantum yield, Φ:

$$\Phi = \frac{\text{number of molecules reacting in a particular process}}{\text{number of photons absorbed}} \qquad (2.10)$$

One of the most useful methods of dosimetry for the determination of quantum yields is the application of chemical *actinometers*. A liquid-phase actinometer containing a specific light-sensitive chemical compound absorbs light of a certain wavelength and gives a product with known quantum yield. The potassium ferric oxalate actinometer works in the 2540–5780 Å range; the uranyl oxalate actinometer, in the 2080–4360 Å range. The chemical reaction is followed in the ferric oxalate actinometer by spectrophotometric determination of the formed ferrous ions, in the uranyl oxalate actinometer by titration of the oxalate concentration or by gas chromatographic determination of the evolved carbon monoxide. Several other chemical actinometers are also known; for example, the so-called Xenometer used in the Xenotest weathering apparatus contains a mercury compound from which Hg forms during irradiation. After measuring the amount of metallic Hg, the mercury compound can be regenerated for repeated applications.

2.4 PRODUCT ANALYSIS METHODS

Some of the reaction products in polymer degradation processes are volatile at ambient temperatures. Among the separation techniques, *gas chromatography* (GC) has received much attention in the past decades; it involves the separation of substances by repeated distribution between a moving gas phase and a solid or liquid stationary phase that are in equilibrium. *Gas-liquid chromatography* (GLC) is the term used when the stationary phase is liquid. This phase is generally spread on an inert solid support which is packed in a column placed in an oven where the temperature can be closely controlled. The mobile carrier gas is usually helium or nitrogen. The gaseous or dissolved sample is injected into the gas stream which carries it into the column. The substances which emerge separately from the column can be detected by various devices, including thermal conductivity and flame ionization detectors. Under well-defined conditions the retention time is characteristic of the various substances. Because of the broad application of GC and GLC in chemistry, we need not deal with further details here. The on-line combination of chromatographic techniques with polymer degradation techniques, in particular *pyrolysis gas chromatography* (PGC), has gained wide acceptance as a method for polymer analysis. Briefly, the pyrolysis of a polymer sample is carried out in a special cell, and the evolved gas is rinsed into the columns by the carrier gas. The pyrolysis can be carried out in furnaces which can be heated very rapidly to the desired temperatures (flash pyrolysis). An interesting pyrolytic cell is used in the Curie-point pyrolizer: the sample is supported by a wire made from a ferromagnetic alloy with a specific Curie point; this is the temperature at which the substance becomes paramagnetic. The wire is placed in a coil fed with high frequency current. An alternating magnetic flux is set up in the wire. Hysteresis losses cause the ferromagnetic wire to become heated. It rapidly reaches the Curie point and maintains this temperature as long as the magnetic transformation is completed. Wires are available with various Curie points, allowing the temperature dependence of pyrolysis to be studied. Two important applications of the PGC method should be mentioned here: (1) with PGC, so-called fingerprints of the polymers can be determined which are unique and readily reproduced, and which allow easy and rapid identification of the material; (2) in addition to NMR and IR techniques, PGC may also be applied to determine monomer sequence distribution in copolymers: appropriate pyrolysis of the sample may produce volatiles characteristic of the various diads or triads.

The combination of gas chromatography and *mass spectrometry* (GC/MS) is an advanced technique in polymer degradation studies: peak identification is facilitated by the combined use of molecular weight and retention data. Very complicated (and expensive) on-line instrumentation has been developed for degradation research purposes. The *Interfaced Pyrolysis Gas Chromatographic Peak Identification System* (IPGCS), developed at the University of Massachusetts, Amherst, incorporates a multipurpose. thermal analyzer allowing slow or ultrarapid temperature rise under inert or reactive atmospheric conditions, various traps, two gas chromatographs, a mass chromatograph, and an infrared vapor-phase spectrophotometer. All are interfaced with a laboratory computer which provides for data acquisition and reduction, and controls the various components. A very similar on-line system has been developed at Princeton University.

In order to illustrate the complexity of polymer pyrolysis products, the mass chromatograms of the volatile products of a low density polyethylene are shown in Figure 2.21. The pyrolysis has been carried out in the Princeton system from room temperature to 600°C, rising by 20°C/min. Products volatilizing between 300 and 600°C were selected for trapping and analysis. In the GC/MS system including two independent gas chromatographs, CO_2 and ClC_2F_5 (Freon 115) carrier gases were chosen because of their highly different molecular weights.

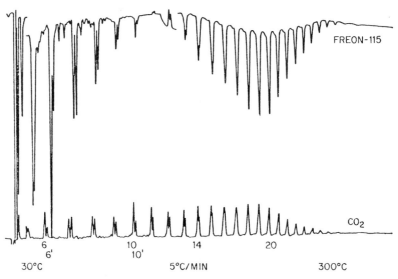

Figure 2.21. Mass chromatograms of the volatile pyrolysis products of low density polyethylene. (Reprinted, by permission of Applied Science Publishers Ltd., from reference 2.11.)

The columns were program-heated at 5°C/min from 30 to 300°C. The assigned structures and the molecular weights for some of the characteristic peaks are: (6) hexene, 84; (6′) hexane, 86; (10) decene, 140; (10′) decane, 142; (14) tetradecene, 196; (20) eicosene (+ eicosane, 283).

A decrease or an increase in average molecular weight generally accompanies degradation processes. *Viscometry* represents a classical method measuring such changes. Although application of *gel permeation chromatography* (GPC) gives more detailed information, the application of viscometry in degradation studies remains important because of its simplicity. A particular advantage of viscometry is that in the case of solution degradation, it can be used for continuous monitoring of the process. In order to follow viscosity changes during degradation, automatic devices have been constructed, for example, to investigate photodegradation kinetics in solution. The apparatus operates on the principle that the solution exposed to irradiation in an irradiation cell is transferred by pressure to the viscometer where time of flow through the capillary is precisely measured by light-sensitive detectors. After measurement, the solution is returned to the cell by applying pressure.

The theory and practice of viscometry and GPC measurements will not be treated here; in connection with the application of GPC in polymer degradation studies, it should be emphasized that the possibilities have not yet been sufficiently exploited. GPC measurements may supply not only average molecular weight data but also information on changes of the molecular weight distribution (MWD) thus providing a better insight into the degradation mechanism. In Figure 2.22 the change of MWD of isotactic polypropylene is shown during thermal oxidation at 130°C under 760 torr and 80 torr O_2 pressure, respectively. At the initial stage of oxidation (even before the onset of measurable oxygen absorption!) the MWD becomes bimodal. With increasing oxidation time, a shift toward higher molecular weights of the peaks with bimodal character can be observed, the shift at 80 torr O_2 pressure being more pronounced than that at 760 torr. Later, the average molecular weights decrease, and MWD becomes narrow and unimodal. As we will see later in connection with polyolefin oxidation, the change of MWD can be interpreted by assuming simultaneous scission and crosslinking processes, the proportion of which is changing during the oxidation process.

In degradation processes in which special products are formed, special methods must be applied. Generally speaking, the application of any investigative method supplying information on the chemical reactions, their products, and chemical or physical changes (i.e., about the mechanism of the degradation) may be useful. In any kind of PVC degradation, HCl evaluation and the formation of conjugated polyenes are the primary processes. Thus the most important (basic) methods of studying PVC degradation involve the quantitative determination of these products. The evolved hydrogen chloride is usually determined by *conducto-*

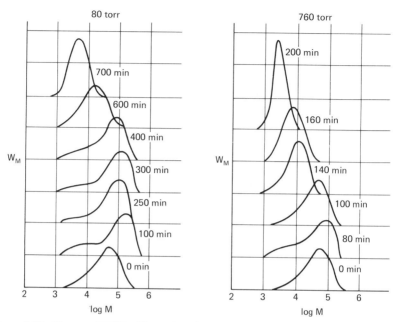

Figure 2.22. Change of MWD of isotactic polypropylene during thermal oxidation at 130°C under 760 and 80 torr oxygen partial pressure, respectively, as measured by gel permeation chromatography. (Reprinted, by permission of John Wiley & Sons, from reference 6.11.)

metry. Equipment for simultaneous dehydrochlorination of six samples is shown in Figure 2.23. The evolved HCl is carried by a neutral rinsing gas from the reaction vessels to conductivity cells containing doubly distilled water, whose conductivity is determined continuously by means of electrodes and a conductivity meter. The PVC samples used in such studies may be powders, films, or dilute solutions (e.g., 1–2% in 1,2,4-trichlorobenzene). Because of the

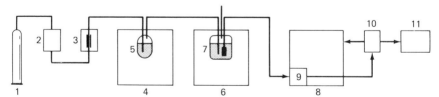

Figure 2.23. Scheme of equipment used for PVC thermal dehydrochlorination studies: (1) gas flask (Ar and N_2) with reductor, (2) pressure buffer, (3) flow meters (6 pieces), (4) oil bath (150–250°C) with temperature regulation, (5) reaction vessels (6 pieces) with samples, (6) water bath (25 or 30°C) with temperature regulation, (7) conductivity cells (6 pieces) with bidistilled water and electrodes, (8) six-point recorder, (9) channel selection, (10) conductivity meter, (11) digital data acquisition system.

very important (catalytic) effect of HCl on the degradation, it must be flushed from the sample thoroughly and immediately; thus the use of solutions or very thin layers of film or powdery samples is necessary.

The polyenes formed during dehydrochlorination of PVC can be determined by *UV-visible spectrophotometry*. In thermal degradation of solution or film samples, the heat treatment can be carried out directly in the spectrophotometer; thus a continuous monitoring of the polyene concentration change is possible. By use of dissolved probes in a separate degradation device such as that shown in Figure 2.23, periodic sampling of the solution for photometric measurements is possible. In the case of powders, the samples must be dissolved in solvents applicable for UV-visible spectrophotometry, such as tetrahydrofuran. The polyenes are very sensitive to oxidation; thus photometric measurement must be carried out immediately after sampling the solvent or dissolving the powder. The change of absorption during oxidation of polyenes at 120°C in an oxygen atmosphere is illustrated in Figure 2.24. In this case, the unoxidized sample had previously been thermally treated in an argon atmosphere at 180°C for 120 min. Extinction coefficients for polyenes of different length have been determined for the film samples and for various solvents; e.g., for trichlorobenzene solutions:

$$\epsilon_m = 10{,}000 + 27{,}700(m - 1) \; [\text{cm}^{-1} \; \ell \; \text{mol}^{-1}] \tag{2.11}$$

where m is the length of the polyene (the number of conjugated double bonds in it).

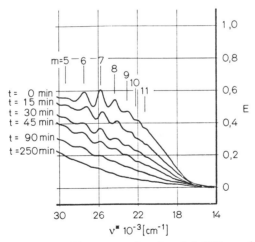

Figure 2.24. Changes in UV-visible spectra of a degraded PVC powder by oxidation at 120°C, 760 torr oxygen. The sample was heat treated for 120 min at 180°C in an argon atmosphere before oxidation. E: extinction (absorbance); ν^*: wave number. (Reprinted, by permission, from reference 2.13.)

2.5 EXTRAPOLATION AND PREDICTION OF SERVICE LIFE

The use of intensified investigative methods to accelerate degradation processes which under application conditions are too slow for practical testing necessitates extrapolation of the results to real conditions. The prediction of service life by extrapolation of data obtained under accelerated testing conditions is a very difficult task; in many cases it is impossible. Not only can the numerical characteristics of the composing processes change, such as values of rate constants, but the basic failure mechanism may also change with decrease of temperature, pressure, dose of irradiation, load in static test, or strain cycle in fatigue test. This is true in particular when the testing is performed above T_g or other transition temperatures and the application takes place below these points: the material changes in physical state.

In spite of these obvious difficulties, accelerated aging tests invariably lead to what appear to be reasonable lifetime predictions when extrapolated to service conditions. Usually the temperature, and often only the temperature, is the intensified variable, although this is at times a poor choice. If temperature is chosen, rate constants or shift factors obtained from experiments in a usually narrow temperature interval are substituted into a model which relates temperature with time. The classical model is represented by the *Arrhenius equation* as it is appled in Eq. (2.4). In the case of a simple process, to predict failure time, t (which is proportional to the reciprocal of the rate constant), the equation can be written in the form:

$$\log t = a + \frac{b}{T} \tag{2.12}$$

where $b = E/R$. Equation (2.12) represents the linear dependence of the logarithm of failure time on the reciprocal value of the absolute temperature $(1/T)$. The *Montsinger equation* (1930) assumes a proportionality to the temperature in °C (θ):

$$\log t = a - b\theta \tag{2.13}$$

The *Yelin equation* (1976) suggests a linear dependence on the logarithm of the absolute temperature $(\log T)$ which was successfully applied to the prediction of polyethylene lifetimes from thermal data:

$$\log t = a - b(\log T) \tag{2.14}$$

Recently, Flynn has proposed a very interesting method to prove the possibility of extrapolation from high temperature measurements to ambient temperature. He recognized that the heating rate (β) of the sample is an important factor in anisothermal measurements and that under practical conditions of application not only the temperature but also the heating rate differs from that applied in the accelerated tests (namely, $\beta \approx 0$ in room temperature applications). At low conversions, i.e., when $(1 - \alpha)^n \approx 1$, Eq. (2.4) can be written in a logarithmic form:

$$\log \frac{d\alpha}{dT} = \log \frac{A}{\beta} - \frac{E}{R} \cdot \frac{1}{T} \tag{2.15}$$

Flynn proposed to make measurements with different heating rates and plot $\log (d\alpha/dT)$ vs $1/T$ representations of the results. If in the quite broad range of heating rates which can be realized in thermal analyses ($6\,^\circ\text{C/min} > \beta > 6.10^{-3}\,^\circ\text{C/min}$) the lines remain parallel, i.e., the activation energy, E, is really independent of the heating rate, then the extrapolation may be justified. On the other hand, if the representations do not produce parallel lines, any thought of extrapolation of kinetic parameters should be forgotten.

Extrapolation from accelerated weatherability tests to real conditions is even more difficult. Much empirical knowledge has been collected about this problem. It is estimated, for example that 100 hours in a specific Weather-Ometer (Type XI Atlas) is equivalent to approximately one year of outdoor exposure. Based on collected results from long exposures, Howard and Gilroy found the following relationship between the two kinds of exposure tests:

$$t_y = (t_h - a)^b \tag{2.16}$$

where t_h is the time in hours of exposure in a Weather-Ometer, t_y is the time in years of exposure outdoors to obtain the same change in the samples, and a and b are specific parameters related to the type of weathering, location, and material involved.

Empirical equations for predicting failure time also exist in the field of mechanical degradation. Although the two main components of a fracture process, crack initiation and crack propagation, obviously differ in their temperature, stress, etc., dependence, prediction of fracture time has been attempted. Zhurkov and Bueche independently found that the time to fracture (t), under uniaxial stress (σ_0), of some polymers below their glass transition temperatures could be expressed by an exponential relationship involving three kinetic parameters:

$$\log t = \log t_0 + \frac{U_0 - \gamma\sigma_0}{T} \tag{2.17}$$

where U_0 was interpreted as activation energy of scission of the chemical bonds, t_0 as the inverse of molecular oscillation frequency, and γ as a structure-sensitive parameter. A comparison of Eqs. (2.12) and (2.17) shows that the *Zhurkov-Bueche equation* describes an Arrhenius-like temperature dependence with an activation energy modified by the applied stress.

REFERENCES

2.1. Alger, R. S. *Electron Paramagnetic Resonance, Technique and Application.* New York: Interscience, 1968.

2.2. Allen, D. W. *Techniques of Polymer Characterization.* London: Butterworths, 1959.

2.3. Bowen, E. J. (ed.). *Luminescence in Chemistry.* London: Van Nostrand, 1968.

2.4. Brandrup, J. and Peebles, L. H., Jr. On the chromophore of polyacrylonitrile. IV. Thermal oxidation of polyacrylonitrile and other nitrile-containing compounds. *Macromolecules* 1:64 (1968).

2.5. Braun, D., Cherdron, H. and Kern, W. *Techniques of Polymer Synthesis and Characterization.* New York: Wiley-Interscience, 1972.

2.6. Grassie, N. and Weir, N. A. The photooxidation of polymers. I. Experimental methods. *J. Appl. Polymer Sci.* 9:963 (1965).

2.7. Hawkins, W. L. Methods for measuring stabilizer effectiveness. In (Hawkins, W. L., ed.) *Polymer Stabilization.* New York: Wiley-Interscience, 1972.

2.8. Howard, J. B. and Gilroy, H. M. Natural and artificial weathering of polyethylene plastics. *Polymer Eng. Sci.* 9:286 (1969).

2.9. Kamal, M. R. (ed.). *Weatherability of Plastic Materials.* New York: Interscience, 1967.

2.10. Ke, B. Newer Methods of polymer characterization. In (Mark, H. F. and Immergut, E. H., eds.) *Polymer Reviews,* Vol. 6. New York: Wiley-Interscience, 1964.

2.11. Kiran, E. and Gillham, J. K. Pyrolysis–molecular weight chromatography–vapour phase infrared spectrophotometry: an on-line system for analysis of polymers. In (Grassie, N., ed.) *Developments in Polymer Degradation–2.* London: Applied Science Publishers, 1979.

2.12. McNeill, I. C. The application of thermal volatilization analysis to studies of polymer degradation. In (Grassie, N., ed.) *Developments in Polymer Degradation–1.* London: Applied Science Publishers, 1977.

2.13. Nagy, T. T., Kelen, T., Turcsányi, B. Tüdős, F. Die Rolle der Polyenoxidation beim thermooxidativen PVC-Abbau. *Angew. Makromol. Chem.* 66:193 (1978).

2.14. Parker, C. A. Phosphorescence and delayed fluorescence from solutions. In (Noyes, W. A., Jr., Hammond, G. S. and Pitts, J. N., Jr., eds.) *Advances in Photochemistry,* Vol. 2. New York: Wiley-Interscience, 1964.

2.15. Rånby, B. and Rabek, J. F. *ESR Spectroscopy in Polymer Research.* Heidelberg: Springer, 1977.

2.16. Schard, M. P. and Russel, C. A. Oxyluminescence of polymers. I. General behavior of polymers. *J. Appl. Polymer Sci.* 8:985 (1964).

2.17. Shalaby, S. W. (ed.). *Thermal Methods in Polymer Analysis.* Philadelphia: The Franklin Institute Press, 1978.

2.18. Still, R. H. The use and abuse of thermal methods of stability assessment. In (Grassie, N., ed.) *Developments in Polymer Degradation–1.* London: Applied Science Publishers, 1977.

Chapter 3
Depolymerization

Depolymerization is essentially a reversal of the polymerization process. In the simplest case, depolymerization consists of initiation at chain ends, depropagation, and termination. In the depropagation step, monomer is unzipped rapidly from the activated chain ends. The main characteristic of degradation processes with dominating depolymerization character is the high monomer yield.

Depolymerization is not the only way in which a polymer can undergo degradation. The type of degradation favored under the given circumstances is the fastest among the possible reactions. Thus, polymethyl methacrylate (PMMA) can depolymerize ($\sim 250°C$), undergo random chain scission ($\sim 300°C$), and lose methanol ($\sim 320°C$). However, we classify PMMA as a depolymerizing material because depolymerization is that degradation process which takes place at the lowest temperature when the temperature is raised continuously as in thermogravimetry.

3.1 CHEMICAL AND THERMODYNAMIC ASPECTS OF DEPOLYMERIZATION

Depolymerization can be the dominating degradation process of a polymer if (A) initiation by main chain scission is possible, (B) the intermediates of the process are stable, and (C) the unzipping of monomers requires a relatively small activation energy.

(A) Initiation by main chain scission is a necessary condition of depolymerization because this process produces the terminal active site capable of depropagation. Although depolymerization can be initiated by random chain scission or by scission at the weak sites built into the main chain, *initiation at the chain end* is typical. This is usually possible if the chain end itself is a weak site of the polymer as in the case of macromolecules with terminal double bond formed by termination with disproportion of the growing radicals:

$$\sim\!\!-CH_2\!-\!\underset{\underset{Y}{|}}{\overset{\overset{X}{|}}{C}}\!-\!CH\!=\!\underset{\underset{Y}{|}}{\overset{\overset{X}{|}}{C}} \longrightarrow \sim\!\!-CH_2\!-\!\underset{\underset{Y}{|}}{\overset{\overset{X}{|}}{C}}\cdot + \left(\dot{C}H\!=\!\underset{\underset{Y}{|}}{\overset{\overset{X}{|}}{C}}\right) \qquad (3.1)$$

The monomer radical formed in this reaction usually volatilizes, but the macro-radical remains in the system as long as its length is higher than the volatilization limit at the degradation temperature, which is generally small as compared to the length of macromolecules. Both X and Y in reaction (3.1) may be hydrogen, but disubstituted polymers (because of the stability requirement) undergo more facile depolymerization.

In initiation by random chain scission, two macroradicals are formed with different terminal groups. Thus the possibility of side reactions distorting the depolymerization character of the process is higher than with chain-end initiation:

$$\sim\!\!-CH_2\!-\!\underset{\underset{Y}{|}}{\overset{\overset{X}{|}}{C}}\!-\!CH_2\!-\!\underset{\underset{Y}{|}}{\overset{\overset{X}{|}}{C}}\!-\!\!\sim \longrightarrow \sim\!\!-CH_2\!-\!\underset{\underset{Y}{|}}{\overset{\overset{X}{|}}{C}}\cdot + \cdot CH_2\!-\!\underset{\underset{Y}{|}}{\overset{\overset{X}{|}}{C}}\!-\!\!\sim \qquad (3.2)$$

The amount of polymer molecules with terminal double bonds and the possibility of chain-end initiation may be decreased by various methods, including the addition of chain transfer agents to the polymerization system.

(B) The terminal radicals formed in the initiation step or in the depropagation steps must be stable enough to *not* participate in various side reactions such as chain transfer. This condition can be fulfilled in various ways: The radical can be stabilized by *resonance stabilization*, as in polystyrene (PS):

$$\sim\!\!\sim\!\!-CH_2\!-\!\underset{\bigcirc}{\overset{\overset{H}{|}}{C}}\cdot \;\rightleftharpoons\; \sim\!\!\sim\!\!-CH_2\!-\!\underset{\bigcirc}{\overset{\overset{H}{|}}{C}} \qquad (3.3)$$

or by means of *steric factors*, as in PMMA. A comparison of the monomer yields in the degradation of PMMA ($X = CH_3$) and PMA ($X = H$) shows the importance of the steric hindrance caused by the α-methyl group:

PMA	PMMA	
radical: $\sim\!-CH_2-\overset{\displaystyle H}{\underset{\displaystyle O=C-OCH_3}{C}}\!\cdot$	$\sim\!-CH_2-\overset{\displaystyle CH_3}{\underset{\displaystyle O=C-OCH_3}{C}}\!\cdot$	(3.4)
monomer yield: $\sim0.7\%$	$>90\%$	

The two stabilizing effects can work together as shown in the comparison of monomer yields of polyethylene, PE ($X = H$, $Y = H$), PS ($X = H$, $Y = C_6H_5$) and poly-α-methylstyrene, PαMS ($X = CH_3$, $Y = C_6H_5$):

PE	PS	PαMS	
radical: $\sim\!-CH_2-\overset{\displaystyle H}{\underset{\displaystyle H}{C}}\!\cdot$	$\sim\!-CH_2-\overset{\displaystyle H}{\underset{\displaystyle C_6H_5}{C}}\!\cdot$	$\sim\!-CH_2-\overset{\displaystyle CH_3}{\underset{\displaystyle C_6H_5}{C}}\!\cdot$	(3.5)
monomer yield: $\sim0\%$	$\sim40\%$	$\sim100\%$	

A very common way of chain transfer is H-abstraction from another macromolecule ($X = H$):

$$\sim\!-\overset{\displaystyle X}{\underset{\displaystyle Y}{C}}-CH_2-\overset{\displaystyle X}{\underset{\displaystyle Y}{C}}\cdot + \sim\!-CH_2-\overset{\displaystyle X}{\underset{\displaystyle Y}{C}}-CH_2\!-\!\sim \;\rightarrow$$

$$\sim\!-\overset{\displaystyle X}{\underset{\displaystyle Y}{C}}-CH_2-\overset{\displaystyle X}{\underset{\displaystyle Y}{C}}-X + \sim\!-CH_2-\overset{\displaystyle X}{\underset{\displaystyle Y}{\overset{\displaystyle \cdot}{C}}}-CH_2\!-\!\sim \quad (3.6)$$

Thus, if the polymer *does not contain active hydrogens*, the probability of chain transfer decreases and that of the depropagation increases. Some examples are shown in (3.4) and (3.5) working together with other effects; an additional example is the comparison of monomer yields of polypropylene, PP ($X = H$) and polyisobutylene, PIB ($X = CH_3$):

PP	PIB	
radical: $\sim\!-CH_2-\overset{\displaystyle H}{\underset{\displaystyle CH_3}{C}}\!\cdot$	$\sim\!-CH_2-\overset{\displaystyle CH_3}{\underset{\displaystyle CH_3}{C}}\!\cdot$	(3.7)
monomer yield: $\sim0.2\%$	$\sim18\%$	

When the above conditions are fulfilled, the terminal radicals will not participate in side reactions but in *depropagation steps* yielding volatile monomer molecules:

$$\sim\!-CH_2-\underset{\underset{Y}{|}}{\overset{\overset{X}{|}}{C}}-CH_2-\underset{\underset{Y}{|}}{\overset{\overset{X}{|}}{C}}-CH_2-\underset{\underset{Y}{|}}{\overset{\overset{X}{|}}{C}}\cdot \longrightarrow \sim\!-CH_2-\underset{\underset{Y}{|}}{\overset{\overset{X}{|}}{C}}-CH_2-\underset{\underset{Y}{|}}{\overset{\overset{X}{|}}{C}}\cdot + \left(CH_2\!=\!\underset{\underset{Y}{|}}{\overset{\overset{X}{|}}{C}}\right) \longrightarrow$$

$$\sim\!-CH_2-\underset{\underset{Y}{|}}{\overset{\overset{X}{|}}{C}}\cdot + \left(CH_2\!=\!\underset{\underset{Y}{|}}{\overset{\overset{X}{|}}{C}}\right) \longrightarrow \text{etc.} \quad (3.8)$$

However, depropagation requires favorable energetic conditions.

(C) The *activation energy* of depropagation is the sum of the activation energy of polymerization propagation and the heat of polymerization. The lower the polymerization heat (i.e., the smaller $|\Delta H_P|$), the lower is the activation energy of depropagation, thus the higher is the probability of depolymerization. While the activation energies of polymerization are of the order of 3–7 kcal/mol, the heats of polymerization are much higher. For monosubstituted monomers ($X = H$), $-\Delta H_P$ ranges from about 16 to 22 kcal/mol; for disubstituted monomers, from about 8 to 13 kcal/mol. For example, in some of the above-mentioned polymers:

MONOMER	$-\Delta H_P$ (kcal/mol)	
Styrene	16.7	
α-Methylstyrene	8.4	(3.9)
Methyl acrylate	18.5	
Methyl methacrylate	13.3	

Depropagation is a β-scission of the terminal radical, producing monomer and a new terminal radical, as shown in reaction (3.8). Splitting off of a hydrogen atom by β-scission,

$$\sim\!-CH_2-\underset{\underset{Y}{|}}{\overset{\overset{X}{|}}{C}}\cdot \longrightarrow \sim\!-CH\!=\!\underset{\underset{Y}{|}}{\overset{\overset{X}{|}}{C}} + H\cdot \quad (3.10)$$

a reaction competitive with depropagation, has a much higher activation energy (by about 10 kcal/mol) than depropagation in cases of disubstituted monomers. In polymers where this difference is not so high, β-scission must be considered as a possible side reaction.

The depropagation process, i.e., the unzipping of monomer molecules, ends either by consuming the macromolecule ($\nu > P_n$) or by a *termination* reaction. Termination can be a uni- or bimolecular reaction, for example, *disproportionation*:

$$\sim\!\!-CH_2-\underset{\underset{Y}{|}}{\overset{\overset{X}{|}}{C}}\cdot + \cdot\underset{\underset{Y}{|}}{\overset{\overset{X}{|}}{C}}-CH_2-\sim \longrightarrow \sim\!\!-CH_2-\underset{\underset{Y}{|}}{\overset{\overset{X}{|}}{C}}-H + \underset{\underset{Y}{|}}{\overset{\overset{X}{|}}{C}}=CH-\sim \qquad (3.11)$$

or *recombination*:

$$\sim\!\!-CH_2-\underset{\underset{Y}{|}}{\overset{\overset{X}{|}}{C}}\cdot + \cdot\underset{\underset{Y}{|}}{\overset{\overset{X}{|}}{C}}-CH_2-\sim \longrightarrow \sim\!\!-CH_2-\underset{\underset{Y}{|}}{\overset{\overset{X}{|}}{C}}-\underset{\underset{Y}{|}}{\overset{\overset{X}{|}}{C}}-CH_2-\sim \qquad (3.12)$$

If the termination products are of high molecular weight, they do not volatilize and can participate in further reactions. For example, the product of reaction (3.11) with unsaturated chain end can react as shown in reaction (3.1).

Because of their very low heats of polymerization, polyaldehydes depolymerize virtually quantitatively to monomer. The depolymerization proceeds in the same manner as in the case of the previously presented vinyl polymers. The degradation of polyaldehydes is governed by the rate of initiation of the depropagation (i.e., the depropagation rate constant is very high). Clearly, any modification of a polymer which would prevent the initiation step of depropagation would stabilize the polymer. Thus, methylation or acetylation of terminal hydroxyl groups of polyformaldehyde raises its degradation temperature by slowing the initiation step of depolymerization. As previously mentioned, copolymers of formaldehyde containing several percent dioxolane are less prone to depolymerization than homopolyformaldehyde because the dioxolane units stop depropagation reaction.

Depending on their structure, polyisocyanates may depolymerize to monomer or form cyclic trimers. Polyisocyanates with aromatic or cyclohexyl substituents are known to depolymerize to monomer when heated to approximately 200°C. Polyisocyanates having short alkyl substituents (R less than five carbons) form cyclic trimers:

$$\left(\underset{\underset{N-CO}{|}}{\overset{R}{|}} \right)_n \rightarrow \quad \underset{\underset{N}{\underset{|}{R}}}{\overset{R\diagdown}{N}} \underset{CO \diagup \diagdown CO}{\overset{CO}{\diagup \diagdown}} \underset{}{N\diagup R}$$

Acetylation of the terminal amino groups of nylon 6 substantially decreases the rate of monomer formation when the polymers are subject to temperatures of about 250°C. This decreased rate of monomer formation suggests that the depropagation step is a back-biting reaction by the amino end group:

$$\sim-NH-\overset{O}{\overset{\|}{C}}-(CH_2)_5-NH_2 \rightarrow \sim-NH_2 + \overset{O}{\overset{\|}{C}}-(CH_2)_5-NH$$

A similar reaction is observed for nylon 7 but only at higher temperatures; this is consistent with the lower thermodynamic stability of eight-membered rings.

The chemical structure of a polymer determines both the *type of degradation* reaction and the temperature at which the degradation reaction will begin.

Let us first consider the thermal degradation of polymers containing nitrile groups. When *polyacrylonitrile* is heated to approximately 260–270°C, degradation begins via a reaction between nitrile groups of the polymer backbone. As shown in Figure 3.1, the reaction between the nitrile groups increases with

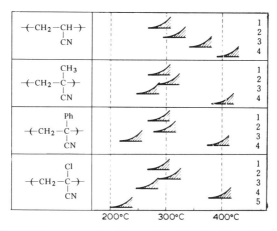

Figure 3.1. Temperature ranges of various reactions for the thermal degradation of polymers containing nitrile groups: (1) reactions of the nitrile groups among each other, (2) random degradation, (3) depolymerization, (4) dehydrocyanation, (5) dehydrochlorination. (Reprinted, by permission of Elsevier Publishing Company, from reference 3.9.)

increasing temperature. Above 300°C, random degradation of the polymer begins to occur. Depolymerization and dehydrocyanation reactions require very high temperatures. If conditions are such that reactions between nitrile groups predominate, the polyacrylonitrile will eventually lose its original properties due to complete carbonization of the polymer.

The introduction of an α-methyl group in acrylonitrile does not affect the reactions between nitrile groups. The α-methyl group does, however, change the mode of degradation of the polymer. The methyl group causes the temperature at which depolymerization occurs to fall below the temperature necessary for nitrile group reaction; thus, *polymethacrylonitrile* depolymerizes upon heating to give virtually quantitative yields of monomer.

The tendency for polymethacrylonitrile to depolymerize to monomer is the result of greatly decreased rate of chain transfer due to a lack of α-hydrogen atoms, the higher rate of depropagation resulting from steric hindrance of the disubstituted polymer and the lower bond dissociation energy of carbon-carbon bonds in the polymer backbone.

Because of increased steric hindrance and increased resonance stabilization of its monomer, *poly-α-phenylacrylonitrile* readily depolymerizes at a temperature even lower than poly-α-methacrylonitrile.

Despite its disubstituted structure, *poly-α-chloroacrylonitrile* behaves differently. As the polymer is heated to near 200°C, dehydrochlorination occurs to produce a polyene. The polyene does not depolymerize, but reactions between the nitrile groups can occur in the polyene as well. The result of nitrile group reactions is a double-stranded polyene. Such materials are quite suitable for carbonization; hence, they are used in the production of "carbon fibers."

As shown in Figure 3.2, the polyacrylates behave somewhat differently from the polymethacrylates. The presence of an α-methyl group in *polymethyl meth-*

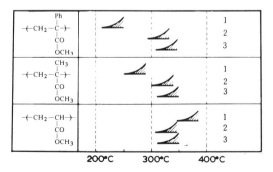

Figure 3.2. Temperature ranges of various reactions for the thermal degradation of polymers containing an ester group: (1) depolymerization, (2) random degradation, (3) methanol formation. (Reprinted, by permission of Elsevier Publishing Company, from reference 3.9.)

acrylate facilitates depolymerization which begins at about 250°C. Similarly, α-phenyl groups enhance depolymerization: *polymethyl α-phenylacrylate* depolymerizes at 220–230°C.

Besides various volatile oligomers, methanol is also found in the degradation products of *polymethyl acrylate*. It is presumed that the ester linkage is stable at the degradation temperature of polymethyl methacrylate but is susceptible to cleavage at the higher degradation temperature (290–300°C) of polymethyl acrylate.

Although this discussion of the various types of degradation implies that they occur as independent processes, the opposite is often quite true. Many reactions can combine (e.g., via radical intermediates) to cause the observed overall degradation process.

3.2 KINETICS OF DEPOLYMERIZATION

Before studying the kinetics of the depolymerization process, first let us define a few important variables of the system and introduce their notation; instead of using concentrations, we will use "number of . . . in the system" because this is more easily applied to solid materials or heterophase systems. Thus:

m = number of monomeric units in the polymers (initial value, m_0)

n = number of polymer molecules (n_0)

m_1 = number of monomers split off ($m_{1,0} = 0, m_1 + m = m_0$)

$s = m_1/n_0$ = *average* number of chain scissions in *one* polymer molecule ($s_0 = 0$)

$P_n = m/n$ = *number average* degree of polymerization ($P_{n,0} = m_0/n_0$)

$\xi = m_1/m_0$ = monomer yield = conversion of depolymerization ($\xi_0 = 0$)

When we deal with the change of the degree of polymerization, we have to distinguish between closed and open systems.

(A) In a closed system, the monomer molecules split off are present and are included when counting the number of molecules; at low conversions when no macromolecules have yet completely depolymerized:

$$P_n = \frac{m + m_1}{n_0 + m_1} = \frac{m_0}{n_0 (1 + s)} = \frac{P_{n,0}}{1 + s} \qquad (3.13)$$

or

$$s = \frac{P_{n,0}}{P_n} - 1 \qquad (3.14)$$

which is the practical definition for the number of chain scissions. However, according to the other definition of s given before

$$s = \frac{m_1}{n_0} = \frac{m_0}{n_0} \cdot \frac{m_1}{m_0} = P_{n,0} \cdot \xi \tag{3.15}$$

From Eqs. (3.14) and (3.15) we have:

$$\xi = \frac{1}{P_n} - \frac{1}{P_{n,0}} \tag{3.16}$$

i.e., the conversion of depolymerization can be calculated directly from the change of the polymerization degree.

(B) In an open system, the monomer molecules split off are not present (they volatilized out of the system); thus at low conversions:

$$P'_n = \frac{m}{n_0} = \frac{m_0 - m_1}{n_0} = \frac{m_0}{n_0}\left(1 - \frac{m_1}{m_0}\right) = P_{n,0}\,(1 - \xi) \tag{3.17}$$

or

$$\frac{P'_n}{P_{n,0}} = 1 - \xi \tag{3.18}$$

Here we used the notation $P'_n = P'_n(t)$ for the degree of polymerization of the depolymerizing polymer in an open system. Equation (3.18) corresponds to the direction b in Figure 1.2, at the beginning of the reaction; it applies when the kinetic chain length of depropagation (v) is smaller than the average degree of polymerization. However, if $v > P_{n,0}$ then each initiation step (let its number be k) causes the complete depolymerization of one macromolecule; then

$$P'_n = \frac{m_0 - k \cdot P_{n,0}}{n_0 - k} = \frac{n_0 \cdot P_{n,0} - k \cdot P_{n,0}}{n_0 - k} = P_{n,0} \tag{3.19}$$

which corresponds to the direction a in Figure 1.2.

The *simplest depolymerization process* consists of three kinds of steps: initiation at the chain end, Eq. (3.1); a series of depropagation steps, Eq. (3.8); and bimolecular termination, Eq. (3.12):

$$
\begin{array}{ll}
\mathrm{P}_i^* \longrightarrow \mathrm{P}_{i-1}^{\cdot} + \mathrm{M} & k_i \\
\hline
\mathrm{P}_{i-1}^{\cdot} \longrightarrow \mathrm{P}_{i-2}^{\cdot} + \mathrm{M} & k_d \\
\mathrm{P}_{i-2}^{\cdot} \longrightarrow \mathrm{P}_{i-3}^{\cdot} + \mathrm{M} & k_d \\
\quad \vdots & \\
\hline
\mathrm{P}_j^{\cdot} + \mathrm{P}_k^{\cdot} \longrightarrow \mathrm{P}_{j+k} & k_t
\end{array}
\tag{3.20}
$$

P_i^* denotes the i-long macromolecules with defective (e.g., double bond containing) chain end (let their number be n_i^*); and $\mathrm{P}_j^{\cdot}$, the j-long radicals (their number, r_j). Let the total number of macromolecules with defective chain end and the total number of radicals be

$$
\Sigma n_i^* = n^* \quad \text{and} \quad \Sigma r_j = r \tag{3.21}
$$

respectively ($r_0 = 0$ and $n_0^* = \alpha n_0$, where α is the fraction of termination by disproportion, characteristic of the polymerization process). By use of these notations, the following differential equations can be written:

$$
\frac{dn^*}{dt} = -k_i n^* \tag{3.22}
$$

$$
\frac{dm_1}{dt} = k_i n^* + k_d r \tag{3.23}
$$

$$
\frac{dr_j}{dt} = k_i n^*_{j+1} + k_d r_{j+1} - k_d r_j - k_t r_j r \tag{3.24}
$$

After summing over all j, we obtain from Eq. (3.24):

$$
\frac{dr}{dt} = k_i n^* - k_t r^2 \tag{3.25}
$$

In the steady state of the reaction, the rate of initiation ($W_i = k_i n^*$) approximately equals the rate of termination ($W_t = k_t r^2$), i.e., $dr/dt \approx 0$. Thus we have from Eq. (3.25):

$$
r \approx \sqrt{\frac{k_i n^*}{k_t}} \tag{3.26}
$$

and, by substituting Eq. (3.26) in Eq. (3.23):

$$\frac{dm_1}{dt} = k_i n^* + k_d \sqrt{\frac{k_i n^*}{k_t}} = k_i n^* \left(1 + \frac{k_d}{\sqrt{k_i k_t n^*}}\right) \tag{3.27}$$

The *kinetic chain length* of the process, v, can be defined as the ratio of the rate of depropagation ($W_d = k_d r$) to the rate of initiation, i.e., it is the number of monomeric units formed as a result of *one* initiation:

$$v = \frac{W_d}{W_i} = \frac{k_d r}{k_i n^*} = \frac{k_d}{k_i n^*} \sqrt{\frac{k_i n^*}{k_t}} = \frac{k_d}{\sqrt{k_i k_t n^*}} \tag{3.28}$$

Substituting Eq. (3.28) in Eq. (3.27) we obtain:

$$\frac{dm_1}{dt} = k_i n^* (1+v) \approx k_i n^* v \approx W_i v \tag{3.29}$$

because $v >> 1$. After dividing by m_0 we obtain the expression for the monomer yield, ξ:

$$\frac{1}{m_0} \frac{dm_1}{dt} = \frac{d\xi}{dt} = k_i \frac{n^*}{m_0} v \tag{3.30}$$

At low conversions $n^* \approx n_0^* = \alpha n_0$ and $v \approx v_0 = k_d / \sqrt{k_i k_t n_0^*}$:

$$\frac{d\xi}{dt} = k_i \alpha \frac{n_0}{m_0} v_0 = k_i \alpha \frac{v_0}{P_{n,0}} \tag{3.31}$$

i.e., the greater the fraction of defective chain ends and the longer the kinetic chains, the higher is the rate of depolymerization. The inverse ratio to the degree of polymerization means that the smaller the macromolecules, the more chain ends are present in the same amount of the polymer.

The integration of Eq. (3.22) leads to:

$$n^* = n_0^* \exp(-k_i t) = \alpha n_0 \exp(-k_i t) \tag{3.32}$$

Substituting Eq. (3.32) in Eq. (3.30), after integration we obtain:

$$\xi = \frac{2\alpha v_0}{P_{n,0}} \left[1 - \exp\left(-\frac{k_i}{2} t\right)\right] \tag{3.33}$$

The time dependence of the monomer yield (and thus also of the number of chain scissions and of the reciprocal value of the number average degree of poly-merization) shows that there is an upper limit of depolymerization and no total conversions can be reached. This is the result of our assumption that initiation is only possible at the defective chain ends. In practice, the depolymerization processes may not be as simplified as described in our treatment.

3.3 CEILING TEMPERATURE

The ceiling temperature (T_c) phenomenon is a thermodynamic concept referring equally to the polymerization and depolymerization processes represented by the general equation

$$M \rightleftharpoons \frac{1}{n}P_n \qquad (3.34)$$

where M represents a monomer molecule, P_n represents a polymer molecule the length of which is the number average degree of polymerization, n. T_c is the temperature at which the partial molar free energy of a monomer molecule (G_M) in a certain polymerization system is equal to the partial molar free energy of a monomer unit in the polymer molecule (G_P) in that system. T_c values are very dependent upon the mechanism of the specific polymerization/depolymer-ization processes and upon the physical states of the monomer and polymer concerned. When referring to T_c values, these conditions should be stated as precisely as possible.

The concept of ceiling temperature is only of direct practical importance under conditions in which the reversible reaction (3.34) is close to its equilibrium position. In practice, most polymerizations and polymer degradations are carried out under conditions far from equilibrium; T_c data are of little impor-tance in these circumstances. The study of T_c phenomena is important in order to obtain information on *thermodynamic* data of polymerization and on the stability of the polymer.

The change in Gibbs free energy of polymerization, ΔG_P, is given by the equation

$$\Delta G_P = G_P - G_M = \Delta H_P - T \Delta S_P \qquad (3.35)$$

where ΔH_P is the heat of polymerization (change of enthalpy), and ΔS_P is the entropy change of polymerization. During polymerization, heat will usually be liberated (exothermic character), and the order in the system increases (the entropy decreases); thus $\Delta G_P < 0$ corresponds to (spontaneous) polymerization and $\Delta G_P > 0$ to (forced) depolymerization. With $\Delta G_P = 0$ we have the polymerization/depolymerization equilibrium at temperature T_c. Then

$$T_c = \frac{\Delta H_P}{\Delta S_P} \qquad (3.36)$$

At temperatures greater than T_c, ΔG_P is a positive quantity showing that reaction (3.34) is thermodynamically impossible in the left to right direction (polymerization); hence the name "ceiling" temperature.

The entropy term in Eq. (3.36) can be expressed as a function of monomer concentration; if the monomer is a gas, the usual standard is 1 atm:

$$\Delta S_P = \Delta S_{P,0} + R \ln p_M \qquad (3.37)$$

where p_M (atm) is the partial pressure of the monomer. If the monomer is in solution, then the standard concentration is 1 mol/ℓ:

$$\Delta S_P = \Delta S_{P,0} + R \ln m \qquad (3.38)$$

where m (mol/ℓ) is the monomer concentration. If the monomer is liquid or solid, then the standard is the pure liquid or solid:

$$\Delta S_P = \Delta S_{P,0} \qquad (3.39)$$

These differences in the entropy term of Eq. (3.36) emphasize the need for accurate specification of the conditions of T_c determination.

The derivation of Eq. (3.36) for T_c involved the basic assumption that G_P is independent of the degree of polymerization. This is strictly valid only when n is infinite. However, for values of $n > 10$-20, the end group effects are normally sufficiently small. A similar assumption can be applied also for the *kinetic derivation* of T_c; here we assume that in the reversible reaction

$$R_i^{\cdot} + M \rightleftharpoons R_{i+1}^{\cdot} \qquad (3.40)$$

the rate constants of polymerization, k_p, and of depolymerization, k_d, are independent of chain length. After summing and using the notation

$$\sum_{i=1} r_i = r \qquad (3.41)$$

we have for the equilibrium:

$$k_p mr = k_d r \qquad (3.42)$$

Thus, using Arrhenius-type temperature dependence:

$$mA_p \exp(-E_p/RT_c) = A_d \exp(-E_d/RT_c) \qquad (3.43)$$

As already mentioned, the activation energy of depolymerization is the sum of the activation energy of polymerization and of the heat of polymerization; thus we have:

$$\frac{A_p}{A_d} m = \exp \frac{\Delta H_P}{RT_c} \qquad (3.44)$$

or

$$T_c = \frac{\Delta H_P}{R \ln \dfrac{A_p}{A_d} + R \ln m} \qquad (3.45)$$

Equation (3.45) is the same as Eq. (3.36) obtained from the thermodynamic derivation for solutions; the ΔS_P term is Eq. (3.36) corresponds then to Eq. (3.38) with

$$\Delta S_{P,0} = R \ln \frac{A_p}{A_d} \qquad (3.46)$$

Table 3.1. Experimentally Observed T_c Values for the Polymerization of Various Monomers.

MONOMER	STATE*	T_c (°C)
Acetaldehyde	ls	−39
Trioxane	lc	36
α-Methylstyrene	lc	45
α-Methylstyrene	ls	61
Tetrahydrofuran	ls	84
Formaldehyde	gc	118
Methyl methacrylate	lc	197
Propylene	gc	271
Styrene	lc	384
Propylene	lc	451
Tetrafluoroethylene	gc	600

*ls = liquid monomer with dissolved polymer; lc = liquid monomer with condensed polymer; gc = gas monomer with condensed polymer.

In Table 3.1 the experimentally observed T_c values for a few selected monomers are collected. In some cases, the effect of polymerization conditions on T_c is also shown.

REFERENCES

3.1. Busfield, W. K. and Merigold, D. The thermodynamics of polymerization of aldehydes and cyclic ethers. I. Formaldehyde and trioxane. *Makromol. Chem.* 138:65 (1970).
3.2. Bywater, S. Photosensitized polymerization of methyl methacrylate in dilute solution above 100°C. *Trans. Faraday Soc.* 51:1267 (1955).
3.3. Dainton, F. S. and Ivin, K. J. Some thermodynamic and kinetic aspects of addition polymerization. *Quart. Rev.* 12:61 (1958).
3.4. Dreyfuss, P. and Dreyfuss, M. P. Polytetrahydrofuran. *Advances in Polymer Science* 4:528 (1967).
3.5. Grassie, N. and McNeill, I. C. Thermal degradation of polymethacrylonitrile. IV. Formation and decomposition of ketene-imine structures. *J. Polymer Sci.* 33:171 (1958).
3.6. Ivin, K. J. Heats and entropies of polymerization, ceiling temperatures, equilibrium monomer concentrations, and polymerizability of heterocyclic compounds. In (Brandrup, J. and Immergut, E. H., eds.) *Polymer Handbook*, 2nd ed. New York: Wiley, 1975.
3.7. Jellinek, H. H. G. Depolymerization. In (Mark, H. F., Gaylord, N. G. and Bikales, N. M., eds.) *Encyclopedia of Polymer Science and Technology*, Vol. 4. New York: Wiley-Interscience, 1966.
3.8. Kern, W. and Cherdron, H. Der Abbau von Polyoxymethylene. *Makromol. Chem.* 40:101 (1960).
3.9. Mita, I. Effect of structure on degradation and stability of polymers. In (Jellinek, H. H. G., ed.) *Aspects of Degradation and Stabilization of Polymers*. Amsterdam: Elsevier, 1978.
3.10. Tobolsky, A. V. and Eisenberg, A. A general treatment of equilibrium polymerization. *Journal of the American Chemical Society* 82:289 (1962).

Chapter 4
Random Chain Scission
and Cross-Linking

The most common form of polymer deterioration is random degradation. In macromolecules, the overwhelming majority of chemical bonds which participate in a specific reaction have practically the same dissociation energy and are statistically equivalent with respect to this reaction. Some degradation processes can only be initiated at the weak sites in the polymer; however, once initiated, the reaction usually proceeds randomly: the active intermediates (e.g., radicals) attack the reactive bonds still present statistically, i.e., without discrimination. The opposite case is also possible; for example, the UV irradiation of PVC causes the formation of some randomly located double bonds. The process will not continue randomly but by the elimination of HCl molecules from neighboring monomer units. Random degradation can be treated on a statistical basis by probability considerations, as well as by formulation of the respective rate equations.

4.1 GENERAL FEATURES

As already mentioned, the most important condition of the randomness of a polymer degradation process is that the participating chemical bonds must be *equivalent*, i.e., they must have about the same bond dissociation energy. Polypropylene contains many kinds of H atoms with various bond dissociation energies, but every monomeric unit of polypropylene contains a tertiary H atom having the lowest $C-H$ bond energy of all. In most of the degradation reactions of polypropylene, these tertiary H atoms will be the reaction sites. The second condition of randomness is the *lack of depolymerization tendency* which has already been discussed in connection with depolymerization.

Typical examples for degradation processes with dominating random chain scission character are hydrolysis, high temperature thermal degradation, and degradation by radiation.   Some condensation polymers, polypeptides, poly-ethers, polyesters, and polysaccharides are inclined to *hydrolytic scission.* When dissolved in strong acids, cellulose is degraded through hydrolytic splitting of the β-glucoside linkages between the structural units of the cellulose chain:

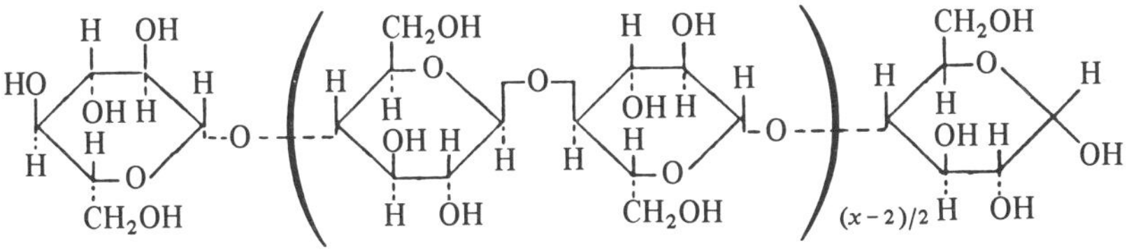

The process can be followed viscometrically, polarimetrically, or by chemical determination of the aldose end group produced for each bond split.   The hydrolysis of nylon 6 (poly-ϵ-caprolactam)

$$NH_2(CH_2)_5 CO \left[NH(CH_2)_5 CO \right]_{x-2} NH(CH_2)_5 COOH$$

in 40% sulfuric acid at 50°C proceeds as a statistical splitting of equivalent bonds. Polydioxolane may be also split by weak acids at either oxygen atom in the main chain:

$$CH_3-CH_2-O-CH_2-O \left[(CH_2)_2 -O-CH_2-O \right]_{x-2} (CH_2)_2 -O-CH_2-OH$$

Not only is acid-catalyzed hydrolysis a random process, but alkaline hydrolysis and alcoholysis of condensation polymers are random processes as well. Studies of the rate of such reactions in homologous series having increasing numbers of monomeric units were among the earliest investigations in the field of reaction kinetics.   The results of these investigations show that the velocity constant, measured under comparable conditions, for the reaction of various members of a given series approaches an asymptotic limit as the chain length increases. In high polymers, i.e., when $P_n \gg 10$, we do not need to distinguish between bonds according to their location in the chain, and end effects can be neglected.

A typical representative of polymers undergoing random *thermal degradation* is polyethylene (PE). The randomness of the process is higher when the number of branches is smaller; polymethylene (prepared in the laboratory from diazo-methane) contains fewer branches than high density PE (prepared by the Ziegler-Natta process) which, in turn, contains fewer branches than low density PE (prepared by high pressure technique). The monomer yield in PE thermal degradation is negligible. The thermal degradation of some condensation polymers and polyalkylene oxides (except polyaldehydes) also proceeds via random chain scission. The degradation temperature of polypropylene oxide ($\sim 275°C$ for a weight loss rate of 0.5%/min) is lower than that of polyethylene oxide ($\sim 325°C$ for the same rate) which, in turn, is lower than those of polypropylene ($\sim 360°C$) and polymethylene ($\sim 390°C$).

Absorption of *high energy radiation* is nonspecific, and all chemical bonds in a polymer have a certain probability of involvement which depends on their electron density. Random cleavage of bonds resulting in fragmentation of macromolecules is typical for degradation by high energy radiation. The process rarely consists of only chain scission; usually cross-linking takes place simultaneously. Whether a polymeric substance consisting of linear macromolecules is converted into a three-dimensional, cross-linked, insoluble, and unmeltable network upon irradiation, or whether it is predominantly degraded in its main chains, depends on the chemical structure of the polymer. As a rule, polymers

$$\sim CH_2 - \overset{\displaystyle X}{\underset{\displaystyle Y}{\overset{\displaystyle |}{\underset{\displaystyle |}{C}}}} \sim$$

of the structure $\sim CH_2 - C \sim$ form a gel, i.e., predominantly cross-link, if X and/or Y are H atoms. Otherwise, main chain scission predominates and the average molecular weight decreases with increasing absorbed dose.

4.2 KINETIC TREATMENT OF CHAIN SCISSION

The randomness of the *bond scission process* means that the probability of cleaveage is proportional to the number of reactive bonds still present. This is the conventional formulation of the rate of unimolecular processes; in this case

$$- \frac{dN}{dt} = kN \tag{4.1}$$

where N is the number of reactive bonds. Let us suppose that those bonds which connect the monomeric units are reactive; then

$$N_0 = P_{n,0} - 1 \tag{4.2}$$

is the initial number of reactive bonds in one molecule. The integration of Eq. (4.1) leads to:

$$N = N_0 \exp(-kt) \tag{4.3}$$

The conversion of bond cleavage is:

$$\xi = \frac{N_0 - N}{N_0} = 1 - \exp(-kt) \tag{4.4}$$

When chain scission takes place with the participation of a reagent, as in the case of hydrolytic bond cleavage, the reaction rate is proportional to the amount of unreacted reagent. Usually, the reagent (here, the acid) is added to the polymer in excess, thus its concentration remains practically constant during the reaction. The process is thus pseudo-unimolecular:

$$k = k'c_0 \tag{4.5}$$

where c_0 is the initial (constant) concentration of the reagent.

The average number of chain scissions in one macromolecule, s, is given by:

$$s = N_0 \xi = (P_{n,0} - 1)\xi \tag{4.6}$$

and the number of macromolecules in the system, n, is

$$n = n_0(1 + s) = n_0[1 + (P_{n,0} - 1)\xi] \tag{4.7}$$

Thus we have the expression for the change of number average degree of polymerization:

$$P_n = \frac{m_0}{n} = \frac{m_0}{n_0(1 + s)} = \frac{P_{n,0}}{1 + (P_{n,0} - 1)\xi} \tag{4.8}$$

assuming that the nubmer of monomeric units in the system (m_0) does not change. A linearized form of this expression using Eq. (4.4) for ξ can be written as follows:

$$\ln \frac{P_n}{P_n - 1} = \ln \frac{P_{n,0}}{P_{n,0} - 1} + kt \tag{4.9}$$

The $\ln P_n/(P_n - 1)$ vs t representation of the polymerization degree data yields a straight line passing through the origin ($P_{n,0} \approx P_{n,0} - 1$). At the beginning of the process $\xi \approx kt$, thus the representation

$$\frac{1}{P_n} = \frac{1}{P_{n,0}} + \xi \cong \frac{1}{P_{n,0}} + kt \tag{4.10}$$

also yields straight lines.

Equation (4.10) is frequently used to obtain the rate constant of the random chain scission process. Figure 4.1 shows the decrease of molecular weight for two types of polystyrene degraded at 280°C. The anionically prepared polystyrene gives a straight line passing through the origin as expected from the theory. On the other hand, for the polystyrene obtained by thermal (radical) polymerization, the line has an appreciable intercept. This shows that chain scission due to weak links occurred at a very early stage of degradation. The weak links detected in this PS sample (0.01–0.02 mol %) are probably incorporated peroxides, but it has been shown that head-to-head structures also lower the thermal stability of PS. As shown in Figure 4.1, the rate in the normal stage for the radical polystyrene is higher than that for the anionic PS, indicating the presence of weak links in the radical PS.

4.3 PROBABILITY TREATMENT OF CHAIN SCISSION

The conversion of bond cleavage, ξ, as defined in Eq. (4.4), is actually a time dependent probability: it is the fraction of the already cleaved bonds. ξ is the probability that a bond randomly selected has undergone scission before time t. Similarly, $1 - \xi$ is the probability that a randomly selected bond is not yet

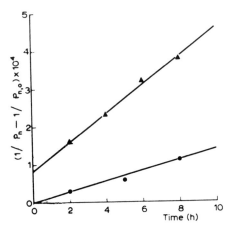

Figure 4.1. Chain scission of polystyrene by thermal degradation at 280°C: ● = anionically prepared sample, $P_{n,0} \approx 5090$; ▲ = thermally initiated radical polymer, $P_{n,0} \approx 4950$. (Reprinted, by permission of Elsevier Publishing Company, from reference 3.9.)

cleaved at time t. Let us now investigate, as Kuhn did it in the 1930s, the question of how many possible ways can be used to split an i-long macro-molecule containing N ($N = i - 1$) reactive bonds in order to produce a P-long ($i > P$) fragment. The answer is simple: using one scission, we can cut off the desired fragment only from the chain ends (two ends = two ways); using two scissions, we can cut out the fragment on $N - P = i - 1 - P$ sites along the chain ($i - 1 - P$ ways). An illustration of this is shown in Figure 4.2, in which the cutting of a molecule consisting of seven units ($i = 7$, $N = 6$) into fragments consisting of three units ($P = 3$) is represented.

Assuming that the process is random and its *events are independent* of each other, we can derive the probability of finding molecules consisting exactly of P units ($P - 1$ bonds) by *multiplying* the probabilities of the individual events which lead to the formation or conservation of such molecules. Since the probability of cleaving *one* bond is ξ, and the probability of it remaining intact is $1 - \xi$, we can derive the following expressions:

(A) For molecules which originally were (and remained) P long:

$$A = n_P(0)(1 - \xi)^{P-1} \tag{4.11}$$

where $n_P(0)$ is the P-th member of the original molecular weight distribution (number of P-long molecules in the original material). It is multiplied by $1 - \xi$ for every conserved bond, i.e., $P - 1$ times.

(B) For P-long molecules formed by splitting off the end of the macromolecules which were longer (at least by one unit) than P:

$$B = 2 \sum_{i=P+1}^{\infty} n_i(0)\xi(1 - \xi)^{P-1} \tag{4.12}$$

The factor 2 is the number of ways of producing such molecules by only *one* cleavage (ξ).

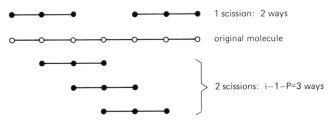

1 scission: 2 ways

original molecule

2 scissions: $i-1-P=3$ ways

Figure 4.2. Possibilities of cutting a molecule consisting of seven units ($i = 7$, $N = 6$) into fragments consisting of three units ($P = 3$) by one or two scisssions.

(C) For P-long molecules cut from the center of macromolecules which were longer (at least by two units) than P:

$$C = \sum_{i=P+2}^{\infty} (i - 1 - P)n_i(0)\xi^2(1 - \xi)^{P-1} \tag{4.13}$$

where $i - 1 - P$ stands for the number of possible ways, and ξ^2 describes the probability, of splitting *two* bonds.

The sum of these three terms is a *distribution function*

$$n_P(\xi) = A + B + C \tag{4.14}$$

which gives the number of P-long molecules in randomly degrading material as a function of conversion, i.e., of time. Since B and C can be reduced to one term with $i = P + 1$ as the lower summation limit ($i - 1 - P = 0$ with $i = P + 1$), we have:

$$n_P(\xi) = n_P(0)(1 - \xi)^{P-1}$$
$$+ \sum_{i=P+1}^{\infty} n_i(0)[2 + (i - 1 - P)\xi]\xi(1 - \xi)^{P-1} \tag{4.15}$$

From the distribution function we are able to determine by summation the various polymerization degree averages. With the well-known definitions being used, the total number of macromolecules in the system is:

$$n(\xi) = \sum_{P=1}^{\infty} n_P(\xi) \tag{4.16}$$

The number average degree of polymerization is:

$$P_n = \frac{1}{n(\xi)} \sum_{P=1}^{\infty} P n_P(\xi) \tag{4.17}$$

The weight average degree of polymerization is:

$$P_w = \frac{1}{n(\xi)P_n} \sum_{P=1}^{\infty} P^2 n_P(\xi) \tag{4.18}$$

Because these averages are easily determined experimentally, the calculation of their values is important if we want to check the validity of the assumptions made about the process. In the following, we will perform only the summation in Eq. (4.16) and compare the result to Eq. (4.7) which was obtained in a very simple way. However, we will present a few additional summation forms which are useful in evaluating Eqs. (4.17) and (4.18).

Substituting Eq. (4.15) into Eq. (4.16), we have an expression

$$n(\xi) = \sum_{P=1}^{\infty} n_P(0)(1 - \xi)^{P-1}$$

$$+ \sum_{P=1}^{\infty} \sum_{i=P+1}^{\infty} n_i(0)[2 + (i - 1 - P)\xi] \xi(1 - \xi)^{P-1} \tag{4.19}$$

which contains a double summation. It can be shown that we may invert the order of summation; for $1 < P < i < m$

$$\sum_{P=1}^{m} \sum_{i=P+1}^{m} f(P, i) = \sum_{i=1}^{m} \sum_{P=1}^{i-1} f(P, i) \tag{4.20}$$

because both kinds of summation include the same terms. For example, with $m = 3$:

P \ i	2	3
1	$f(1, 2)$	$f(1, 3)$
2	$f(2, 3)$	—
3	—	—

i \ P	1	2
1	—	—
2	$f(1, 2)$	—
3	$f(1, 3)$	$f(2, 3)$

The right-hand side of Eq. (4.20) can split into two terms:

$$\sum_{i=1}^{m} \sum_{P=1}^{i-1} f(P, i) = \sum_{i=1}^{m} \left(\sum_{P=1}^{m} f(P, i) - \sum_{P=i}^{m} f(P, i) \right) \tag{4.21}$$

Applying Eqs. (4.20) and (4.21) to Eq. (4.19), with $m = \infty$, we obtain the following expression:

$$n(\xi) = \sum_{P=1}^{\infty} n_P(0)(1 - \xi)^{P-1}$$

$$+ \sum_{i=1}^{\infty} \left\{ n_i(0)[2 + (i-1)\xi]\xi \left(\sum_{P=1}^{\infty} (1-\xi)^{P-1} - \sum_{P=i}^{\infty} (1-\xi)^{P-1} \right) \right.$$

$$\left. - n_i(0)\xi^2 \left(\sum_{P=1}^{\infty} P(1-\xi)^{P-1} - \sum_{P=i}^{\infty} P(1-\xi)^{P-1} \right) \right\} \tag{4.22}$$

The following summation forms ($q < 1$) can be used in evaluating Eq. (4.22) and in calculating the various polymerization degree averages:

$$\sum_{j=k}^{\infty} q^j = \frac{q^k}{1-q} \tag{4.23}$$

$$\sum_{j=k}^{\infty} jq^j = \frac{q^k}{(1-q)^2} [k - (k-1)q] \tag{4.24}$$

$$\sum_{j=k}^{\infty} j^2 q^j = \frac{q^k}{(1-q)^3} [k^2 - (2k^2 - 2k - 1)q + (k-1)^2 q^2] \tag{4.25}$$

$$\sum_{j=k}^{\infty} j^3 q^j = \frac{q^k}{(1-q)^4} [k^3 - (3k^3 - 3k^2 - 3k - 1)q$$

$$+ (3k^3 - 6k^2 + 4)q^2 - (k-1)^3 q^3] \tag{4.26}$$

Using Eqs. (4.23) and (4.24) with the corresponding limits and substituting in Eq. (4.22), we obtain:

$$n(\xi) = \sum_{P=1}^{\infty} n_P(0)(1-\xi)^{P-1} + \sum_{i=1}^{\infty} \left\{ n_i(0)[2 + (i-1)\xi]\xi \left(\frac{1}{\xi} - \frac{(1-\xi)^{i-1}}{\xi} \right) \right.$$

$$\left. - n_i(0)\xi^2 \left(\frac{1}{\xi^2} - \frac{(1-\xi)^{i-1}}{\xi^2} [1 + (i-1)\xi] \right) \right\}$$

$$= \sum_{P=1}^{\infty} n_P(0)(1-\xi)^{P-1} + \sum_{i=1}^{\infty} \{ n_i(0)[1 + (i-1)\xi] - n_i(0)(1-\xi)^{i-1} \}$$

$$= \sum_{i=1}^{\infty} n_i(0)[1 + (i-1)\xi] = n_0 + (m_0 - n_0)\xi$$

$$= n_0[1 + (P_{n,0} - 1)\xi] \tag{4.27}$$

The result derived from the distribution function is in complete agreement with Eq. (4.7).

As we have seen, the number average degree of polymerization after random degradation does not depend on the initial molecular weight distribution (MWD). This is not the case with other characteristics of the distribution, e.g., with the weight average degree of polymerization (P_w). If the original polymer has a random MWD we obtain very simple expressions for random chain scission processes: the new distribution will also be random.

Random MWD's are observed in many kinds of polymers obtained in practical polymerization techniques, e.g., in condensation polymers, in radically polymerized polymers terminated by disproportionation, and also in cationically polymerized materials. In the random distribution (which is often called "Most Probable Distribution"), the number distribution function is

$$n_P = n(1 - \alpha)\alpha^{P-1} \tag{4.28}$$

where n is the total number of macromolecules:

$$\sum_{P=1}^{\infty} n_P = n \tag{4.29}$$

according to the definition. The polymerization degree averages of a random MWD are:

$$P_n = \frac{1}{1 - \alpha} \qquad P_w = \frac{1 + \alpha}{1 - \alpha} \qquad \frac{P_w}{P_n} = 1 + \alpha \tag{4.30}$$

When P_n is high, $\alpha \approx 1$; thus for random distribution polymers, the polydispersity, characteristic for the width of a MWD, is $P_w/P_n \approx 2$.

Substituting Eq. (4.28) for the original distribution in Eq. (4.15), we obtain:

$$n_P(\xi) = n(\xi)(1 - \beta)\beta^{P-1} \tag{4.31}$$

where $n(\xi)$ stands for the total number of polymers according to Eq. (4.27), and β is the distribution parameter of the new MWD:

$$\beta = \alpha(1 - \xi) = 1 - \frac{1 + s}{P_{n,0}} = 1 - \frac{1}{P_n}$$

i.e., β is derived from the distribution parameter of the original MWD, $\alpha = 1 - 1/P_{n,0}$, and from ξ, the conversion of the random scission process, $\xi = s/(P_{n,0} - 1)$. The weight distribution function, W_p, of a polymer with random MWD, normalized to 1 g of polymer (i.e., $\Sigma W_p = 1$) is the following:

$$W_p = P(1 - \beta)^2 \beta^{P-1} \qquad (4.32)$$

In Figure 4.3 the change of the weight distribution of a polymer during random chain scission is illustrated. The original polymer ($s = 0$) had a random distribution ($P_{n,0} = 1000$). (For better comparison, the curves in b are related to their maximum value.) As can be seen, with increasing number of scissions, the average molecular weight decreases, but the distribution remains random.

4.4 RANDOM CROSS-LINKING; SIMULTANEOUS CHAIN SCISSION AND CROSS-LINKING

There are many polymers in which cross-links are produced during degradation. Cross-linking is especially prevalent during radiation degradation; thus the theory has been developed mainly in connection with radiation cross-linking (Charlesby, Saito, Dole, etc.). Cross-linking during degradation has also been studied in connection with the problems of network formation and gelation by condensation reactions and vulcanization (Flory, Stockmayer, etc.).

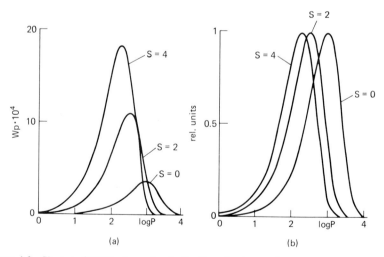

Figure 4.3. Change of MWD of a randomly distributed polymer ($P_{n,0} = 1000$) during random chain scission: (a) weight distribution with $\Sigma W_p = 1$; (b) weight distribution related to the maximum value ($W_{p,max} = 1$). s: average number of scissions in one original macromolecule.

When both cross-links and main chain scissions are produced during the degraduation, it may be assumed that they are results of independent events. We are able to obtain the overall change by considering that cross-linking occurs after all chain scissions have taken place.

Let c denote the number of cross-links related to *one* original macromolecule (as s is the number of scissions per original macromolecule), then the number of macromolecules in the system is given by the expression:

$$n = n_0(1 + s - c) \tag{4.33}$$

Since the number of monomer units does not change, the number average degree of polymerization is:

$$P_n = \frac{m_0}{n} = \frac{m_0}{n_0(1 + s - c)} = \frac{P_{n,0}}{1 + s - c} \tag{4.34}$$

or

$$\frac{P_{n,0}}{P_n} = 1 + s - c \tag{4.35}$$

It can be shown that for polymers originally having random MWD, the weight average degree of polymerization after simultaneous scission and cross-linking can be given as follows:

$$\frac{P_{w,0}}{P_w} = 1 + s - 4c \tag{4.36}$$

As can be seen from the above equations, with increasing amount of cross-linking the molecular weight of the polymer increases, the weight average increasing faster than the number average. As the process continues, a part of the polymer becomes insoluble in any solvent. This insoluble material is called a *gel*, and the process is called *gelation*; the instant at which an incipient gel is created is called the *gel point*. The c value at the gel point, c_g, is given from Eq. (4.36); this is when P_w becomes infinitely large, i.e., when $1/P_w = 0$:

$$c_g = \frac{1 + s_g}{4} \tag{4.37}$$

where s_g is the value of s at the gel point. In the post-gel point period of the degradation, the amount of gel (its weight fraction is denoted by γ) increases and the amount of soluble molecules decreases (the sol fraction, $\sigma = 1 - \gamma$). The relationship between these amounts and the number of cross-links and scissions depends on the initial MWD; for a randomly distributed original polymer, the Charlesby-Pinner relationship is valid:

$$\sigma + \sqrt{\sigma} = \frac{s}{2c} + \frac{1}{2c} \tag{4.38}$$

A plot of $\sigma + \sqrt{\sigma}$ vs $1/c$, the so-called *Charlesby-Pinner plot*, gives a straight line which passes through the origin when only cross-linking occurs. The plot intersects the ordinate when both scission and cross-linking take place. A deviation from linearity is usually caused by deviation of the initial MWD from the most probable distribution.

At the gel point $\gamma = 0$ and $\sigma = 1$, thus $\sigma + \sqrt{\sigma} = 2$. Substituting this value into Eq. (4.38) gives the same c_g as Eq. (4.37). In radiation degradation, both s and c are proportional to the radiation dose, r, and the ratio s/c is equal to a constant, λ. It is convenient in such cases to rewrite Eq. (4.38) in terms of dimensionless parameters:

$$\sigma + \sqrt{\sigma} = \frac{\lambda}{2} + \left(2 - \frac{\lambda}{2}\right)\frac{r_g}{r} \tag{4.39}$$

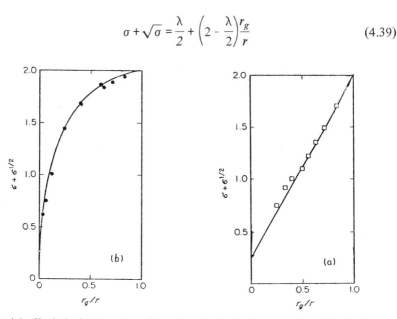

Figure 4.4. Charlesby-Pinner plot of irradiated polyethylene samples: (a) fractionated ($P_n \approx 37,000$); (b) unfractionated. (Reprinted by permission from *The Radiation Chemistry of Macromolecules*, Vol. I., p. 246, M. Dole, Editor, Academic Press, N.Y. 1972.)

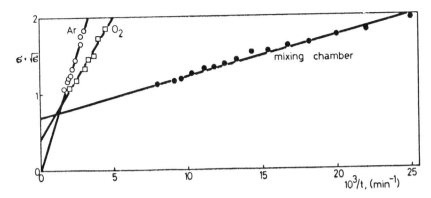

Figure 4.5. Charlesby-Pinner plot of gel formation during PVC degradation at 180°C, under neutral (Ar) and oxidative (O_2) conditions, and in the mixing chamber of a torque rheometer. (Reprinted, by permission, from reference 5.15.)

where r_g is the radiation dose at the gel point ($c_g/c \approx r_g/r$): during the post–gel point irradiation, r_g/r decreases from 1 to 0 with increasing dose. Charlesby-Pinner plots of irradiated polyethylene samples are shown in Figure 4.4. Obviously, the initial MWD of the nonfractionated sample was not random.

During thermal and thermo-oxidative degradation of PVC not only do HCl elimination and polyene formation take place as is often assumed, but secondary processes such as benzene formation and cross-linking also occur. Charlesby-Pinner plots of data on gels formed during degradation under different conditions are shown in Figure 4.5. Under an inert atmosphere only cross-linking occurs, while under an oxygen atmosphere or under dynamic conditions both scission and cross-linking take place. In Figure 4.5, a representation in function of $1/t$ was used because the rate of cross-linking was found to be constant ($c = kt$).

REFERENCES

4.1. Blazsó, M. and Székely, T. Thermal degradation of polyvinyl-trimethyl-silane by pyrolysis gas chromatography. *European Polymer J.* 10:115 (1974).

4.2. Charlesby, A. *Atomic Radiation and Polymers.* New York: Pergamon Press, 1960.

4.3. Dole, M. (ed.). *The Radiation Chemistry of Macromolecules*, Vols. 1 and 2. New York: Academic Press, 1972.

4.4. Flory, P. J. Molecular size distribution in three dimensional polymers. *Journal of the American Chemical Society* 63:3083, 3091, 3096 (1941).

4.5. Kuhn, W. Über die Kinetik des Abbaues hochmolekularer Ketten *Berichte Deutschen Chem. Ges.* 63:1503 (1930).

4.6. O'Donnel, J. H., Rahman, N. P., Smith, C. A. and Winzor, D. J. Chain scission and cross-linking in the radiation degradation of polymers: limitations on the utilization of theoretical expressions and experimental results in the Pregel region. *Macromolecules* **12**:113 (1979).

4.7. Saito, O. Statistical aspects of infinite network formations. *Polymer Engineering and Science* **19**:234 (1979).

4.8. Simha, R. Kinetics, degradation and size distribution of long chain polymers. *Journal of Applied Physics* **12**:569 (1941).

4.9. Stockmayer, W. H. Theory of molecular size distribution and gel formation in branched polymers. *Journal of Chemical Physics* **11**:45 (1943); **12**:125 (1944).

Chapter 5
Degradation Without
Chain Scission

A very important type of polymer degradation consists of those processes which take place without scission of the main polymer chain; characteristic of this kind of degradation is the *participation of side groups* in the reaction. Because the rules governing these processes are very similar to those which determine the reactions of the corresponding low molecular weight compounds containing the same reactive groups, and the differences take their origin only from the macromolecular nature of the material, these processes are often called *polymer-analogous reactions*. There are two main differences between the polymeric material and its model compounds: (1) the dissociation energy of a given bond has much greater diversity because of the variety of the polymer surroundings, and is usually smaller than in the low molecular weight model; (2) the side groups in the polymers are fixed in their position, and this topological arrangement allows the development of special kinds of chain reactions (e.g., unzipping) which are not possible in the low molecular weight equivalent.

A systematic and comprehensive discussion of side chain degradation cannot be undertaken here because so many different reactions are involved corresponding to the wide variety of possible side groups. Only some of these will be treated. Emphasis will be given to the degradation of polyvinyl chloride, partly because of the importance of this polymer and partly because of the author's personal research in the field of PVC degradation.

5.1 GENERAL REMARKS

From vinyl polymers of the structure $+CH_2-CHX+_n$ where X is an electronegative group (e.g., a halogen atom, or a hydroxyl or acetate group), HX elimination is the most common form of degradation:

$$\sim-CH_2-\underset{\underset{X}{|}}{CH}-\sim \;\rightarrow\; \sim-CH=CH-\sim + HX \qquad (5.1)$$

The elimination of side groups from the polymer, like the elimination of HX from low molecular weight compounds, usually proceeds via a *nonradical* mechanism. These *unimolecular* (in the case of polymers, intramolecular) HX elimination reactions have a frequency factor of the order of 10^{13} sec^{-1} and an activation energy of 40–60 kcal/mol. As shown in Table 5.1, activation energies of HX elimination are substantially lower than the dissociation energies of C—X bonds. H_2O elimination has the highest activation energy and, surprisingly, acetic acid elimination, the lowest. This can be explained by the possibility of the formation of a six-membered ring in the transition state of acetic acid elimination:

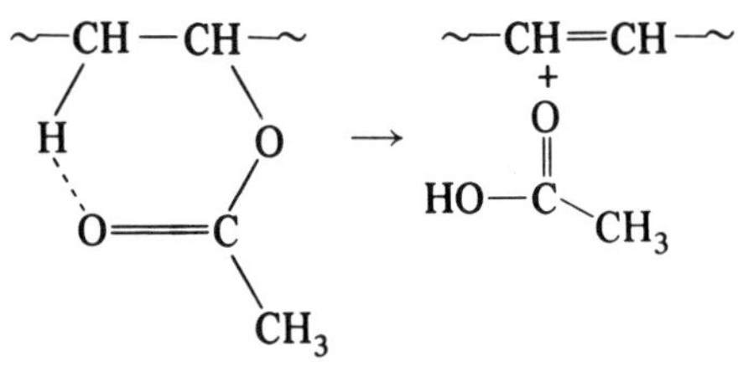

The formation of a favored six-member transition state may explain the deviation from the elimination rate sequence expected from considerations of the se-

Table 5.1 Activation Energies of the Elimination of HX from RX in Gas-phase Reaction (kcal/mol). The Figures in Parentheses Are the Dissociation Energies of R—X Bonds.

R \ X	Cl	Br	OCOCH$_3$	OH
Ethyl	60.8 (81)	53.9 (69)	47.8 (85)	–
i-Propyl	50.5 (81)	47.8 (68)	45.0	–
t-Butyl	45.0 (79)	41.0 (63)	40.0	65.5 (93)

quence of electronegativities and polarizabilities: $Br- > Cl- > > CH_3COO- >$ $HO-$.

The activation energy of HCl or HBr elimination at tertiary halogen atoms is about 5-6 kcal/mol lower than at secondary halogen atoms (see Table 5.1); correspondingly, the elimination rate at a tertiary halogen atom is about 100 times greater than the rate of elimination at the regular secondary halogen atoms. Thus, irregularities of the structure (e.g., branching points) have important influence on HX elimination. Once a *double bond* is formed by the first elimination step at such a "weak site," it represents a weak site in itself and the next elimination step is again an "activated" elimination, leading to the formation of a *conjugated double bond*, and so forth. The process is a special kind of chain reaction in which the propagation proceeds intramolecularly. This specific character of elimination from polymers is responsible for the higher rate and much lower heat stability compared to the low molecular weight model compounds. When the possibility of intramolecular propagations is precluded, i.e., in polymers containing isolated halogen atoms (e.g., chloroprene rubbers or copolymers of vinyl chloride), HX elimination proceeds substantially more slowly.

A typical representative of a polymer which decomposes by elimination of the side group is *polyvinyl chloride* (PVC). PVC degradation has been discussed in sections 5.2-5.4. It should simply be mentioned here that the radical, ionic, or molecular character of HCl elimination from PVC is one of the most disputed questions of polymer degradation. Evidence exists to support each of these mechanisms. The fact that in the presence of labeled toluene, radioactivity is incorporated into the PVC during dehydrochlorination is regarded as evidence of a radical mechanism. Some other experimental observations, especially in the field of PVC stabilization, can only be explained by an ionic degradation mechanism. It can be assumed that the experimental conditions strongly influence the course of degradation and the contribution of the various mechanisms depends on such factors as the character of the medium, the presence of oxygen, etc.

The degradation of *polyacrylonitrile* (PAN) is interesting because an intramolecular chain reaction with side group participation is possible not only by HCN elimination but also via a reaction between nitrile groups ("polymerization"). Propionitrile, a model compound for PAN, eliminates HCN with an activation energy of 77 kcal/mol during gas-phase pyrolysis. Accordingly, dehydrocyanation of PAN should occur at about 350°C. However, as shown in Figure 3.1, nitrile group polymerization begins at about 270°C, then random degradation and depolymerization proceed. HCN elimination begins only at about 400°C.

The side group coupling in PAN has a relatively short sequence: the conjugation in the $-\overset{|}{C}=N-\overset{|}{C}=N-$ side chain extends only to 5-6 units. The $-C\equiv N$ group polymerization is interrupted by various transfer reactions:

intramolecular transfer

intermolecular transfer

β-scission

(5.3)

The radical on the main chain, formed by intramolecular chain transfer to an α-hydrogen atom, may undergo β-scission leading to a certain weight loss. However, in the presence of polar groups such as acrylates (introduced into PAN, for example, by copolymerization with acrylic acid), the $-C\equiv N$ side chain polymerization proceeds via an ionic mechanism, and β-scission (along with weight loss) is significantly reduced.

The possibility that side group reactions other than HX elimination may be the dominant degradation route is demonstrated in *polyacrylates* and *polymethacrylates*. In many cases, volatile olefins are formed from the alcohol part of the ester, and the acid fragment remains attached to the main chain. For example, isobutylene forms during the degradation of poly-*t*-butyl methacrylate:

(5.4)

The rate of this reaction is higher than expected from decomposition of model compounds. This can be caused by the accelerating effect of the acid formed on the reaction of the ester group in the next monomeric unit.

When primary alkyl groups (such as methyl, ethyl, or *n*-butyl) form the ester moiety in a polyacrylate or polymethacrylate, main chain scission becomes more important than olefin formation (for methyl esters see Figure 3.2). Polymethacrylates such as polymethyl methacrylate almost completely depolymerize. The acrylates undergo random scission rather than depolymerization. In addition to the chain scission products, alcohols are also formed. Alcohol formation, which is as high as 50-60% in the case of methyl or benzyl esters, cannot be explained by the usual acyl-oxygen bond scission:

$$
\begin{array}{cc}
-\text{CH}- & -\text{CH}- \\
| & | \\
\text{C}=\text{O} & \text{C}=\text{O} \\
--\overset{|}{+}-- \rightarrow & \cdot \\
\text{O} & + \\
| & \\
\text{CH}_3 & \text{CH}_3\text{O}\cdot
\end{array}
\qquad (5.5)
$$

in which the formed alkoxy radical would abstract a hydrogen atom. Alcohol elimination rather seems to follow an intramolecular route:

$$
\begin{array}{ccc}
\diagdown \text{C} \diagup & & \diagdown \text{C} \diagup \\
& \rightarrow & \| \\
\text{H} \quad \text{C}=\text{O} & \rightarrow & \text{C} \quad + \text{CH}_3\text{OH} \\
\diagdown \text{O} \diagup & & \| \\
| & & \text{O} \\
\text{CH}_3 & &
\end{array}
\qquad (5.6)
$$

where a ketene and methanol are formed. This behavior is similar to that observed in the pyrolysis of acetic acid.

Reactions of the polymer side groups are important not only with respect to degradation but also for intentional alteration of some polymers. Thus, for example, the industrial preparation of polyvinyl alcohol from polyvinyl acetate is carried out by alcoholysis of the acetate groups by methanol. Similarly, many kinds of side group transformations of various cellulose derivatives are of practical importance.

5.2 THERMAL DEGRADATION OF PVC

Polyvinyl chloride is produced on a very large scale. It is an important polymer, having numerous advantageous properties. However, it has the undesirable characteristic of being very unstable. The thermal degradation of PVC begins at relatively low temperatures; spontaneous dehydrochlorination is possible at temperatures as low as $150°C$:

$$
\sim(\text{CH}_2-\underset{\underset{\text{Cl}}{|}}{\text{CH}})_n\sim \rightarrow \sim(\text{CH}=\text{CH})_n\sim + n\text{HCl} \qquad (5.7)
$$

In the course of this process, polyenes of various lengths are formed which cause observable discoloration even at very low concentrations. Thus, PVC degradation has a harmful effect even if it occurs to only a small extent. Degradation not only causes discoloration but changes other polymer properties as well.

The degradation usually does not consist only of the primary processes represented in reaction (5.7); secondary reactions must also be taken into account.

For example, polyenes can undergo a variety of thermal transformations such as cyclization and cross-linking. Dehydrochlorination may be catalyzed by the evolved HCl. Various modes of chain scission are possible. Clearly, the degradation of PVC is an extremely complex process. Extensive research (theoretical and applied) in this field has been done for several decades. Although substantial progress has been made, especially with respect to the stabilization of the polymer, advances have often been made by empirical means; consequently, the basic details of the mechanism of PVC degradation are still unknown. Many explanations of PVC degradation have begun to converge on a general scheme. It has recently been demonstrated that results which do not agree with the theory were obtained under conditions differing greatly from those used in the majority of studies. When experimental conditions are unambiguously fixed, results can be uniformly interpreted. At present, PVC degradation can be explained by a fairly consistent (and widely accepted) mechanism.

Primary Processes

In the dehydrochlorination of PVC solutions — particularly at relatively low temperatures — three different stages can be distinguished in the time dependence of the conversion measured by conductometry (ξ_{HCl}). At the beginning of the process, an induction period can be observed, the length of which decreases with increasing temperature, increasing concentration of polymer solution, and increasing flow rate of argon gas. (We would remark that an induction period — though generally shorter — can also be observed in measurements made with films and powder samples.) The rate measured after the induction period decreases with increasing conversion and becomes practically constant.

The coloration of PVC resulting from thermal degradation is the consequence of several absorption bands appearing in the near UV and in the visible range of the spectrum. Although each band is formed by the superposition of the spectra of several polyenes, the extinction at the maxima can to a good approximation be related to the extinction of polyenes of a given length. In this way, the concentration of polyenes of various lengths can be estimated from the time dependence of extinction. The conversion calculated on this basis, ξ_p, shows a picture similar to the kinetics of the dehydrochlorination process, but without any induction period (Figure 5.1).

As shown in Figure 5.1, two different rates can be determined from the experimental data, namely, a higher initial rate, W_0, and a lower steady rate, W_{st}. The latter can be used for the determination of the apparent "initial" conversion (ξ_0), after extrapolation to time 0. The rate W_0 does not give a straight line in the Arrhenius plot. This is due to the different activating effects of the various structural defects and will be discussed later.

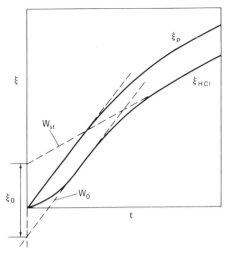

Figure 5.1. Schematic representation of typical conversion curves of PVC degradation in solution, obtained from conductometric (dehydrochlorination, ξ_{HCl}) and spectrophotometric (polyene formation, ξ_p) measurements.

The hydrogen chloride formed is at first present in dissolved form, then desorbs into the gas phase, and is only subsequently transported into the conductivity cell. As the experimental control of the process is carried out by conductivity measurements, we can determine experimentally only the value of the HCl transported out of the system. Thus, the conversion rate is in agreement with the rate determined only after reaching steady state: hydrogen chloride then remains in the solution only in a quantity corresponding to the dynamic equilibrium, and the total amount of hydrogen chloride formed is transported to the conductivity cell. The amount of dissolved hydrogen chloride (as well as the length of the induction period t_i) depends on parameters which determine reaction rates, desorption, and transport.

The above interpretation of the *induction period* is in agreement with the fact that the formation of polyenes is initiated at the very beginning of the process, and no change can be observed in the formation rate of polyenes at the end of the induction period. The generally shorter induction period observed on films or powder samples can be similarly interpreted, i.e., by the retention of hydrogen chloride until dynamic equilibrium is reached.

The overall decomposition reaction can be regarded as a special unzipping reaction in which the initiating step is predominantly a random decomposition of monomeric units, accompanied by initiation of a higher rate occurring at the structural defects of the polymer. The initial overall rate develops as a superposition of decompositions arising from both kinds of initiation, while later, after consumption of defects, the lower overall rate represents a randomly initiated decomposition.

Propagation takes place in a series of activated unimolecular steps at a rate higher by orders of magnitude than the rate of random initiation, as shown by experiments on model compounds. The kinetic chain length is short: the polyenes formed involve an average of only about six double bonds.

It is difficult to understand why chain propagation stops at such an early stage. Therefore, the crucial point of every correct model of PVC degradation is a physically reliable interpretation of chain termination. We may assume several possibilities: for example, it can be imagined that the series of allyl-activated steps reaches the end of a polymer segment which is intact from the kinetic point of view, or that a reaction step interrupts the propagation of activation. Secondary reactions leading to termination, e.g., cyclization via backbiting or branching, are also probable.

In the following material, we will present a simplified kinetic treatment of the above model, in which chain termination is not formulated in a separate reaction but is taken into account by a probability (stoichiometric) parameter δ. In the course of hydrogen chloride elimination from a monomeric unit (M), an activating double bond (P*) and an inactive product (P) can be formed with a given probability. The overall process can be expressed by the following scheme:

random initiation (rate constant k_1):

$$M \longrightarrow (1 - \delta)P_1^* + \delta P_1 + HCl \tag{5.8}$$

initiation at the weak sites (k_1^*):

$$H(+M) \longrightarrow (1-\delta)P_1^* + \delta P_1 + HCl \tag{5.9}$$

propagation (unzipping, k_2):

$$\begin{aligned}
P_1^*(+M) &\longrightarrow (1-\delta)P_2^* + \delta P_2 + HCl \\
P_2^*(+M) &\longrightarrow (1-\delta)P_3^* + \delta P_3 + HCl \\
&\,\cdots \\
P_i^*(+M) &\longrightarrow (1-\delta)P_{i+1}^* + \delta P_{i+1} + HCl
\end{aligned} \tag{5.10}$$

We put M into parentheses in Eqs. (5.9) and (5.10) to demonstrate that despite the fact that the reactions are unimolecular and that their rates are proportional to the concentration of defects (H) or of activating double bonds (P*), a neighboring monomeric unit decomposes in the course of the reaction. In the course of the chain propagation steps the activating double bond (e.g., in P_i^* containing i double bonds, the i-th) becomes "inactivated", and either an activating double bond (P_{i+1}^*, in $1 - \delta$ probability) or an inactive product (P_{i+1}, a polyene containing $i + 1$ double bonds, in δ probability) is formed as a consequence of the decomposition of the neighboring monomeric unit.

If p_i represents the concentration of i-long polymers and M_0 the initial concentration of monomeric units, and if the conversion is given as $\xi = HCl/M_0$, then on the basis of this mechanism the following rate equation can be written:

$$\frac{d\xi}{dt} = W_{in} + k_2(p_1^* + p_2^* + \ldots + p_i^* + \ldots) = W_{in} + k_2 \sum_{i=1}^{\infty} p_i^* \qquad (5.11)$$

which describes the change of conversion. For the active polyenes we have:

$$\frac{dp_1^*}{dt} = (1-\delta)W_{in} - k_2 p_1^*$$

$$\frac{dp_2^*}{dt} = (1-\delta)k_2 p_1^* - k_2 p_2^* \qquad (5.12)$$
$$\ldots$$

$$\frac{dp_i^*}{dt} = (1-\delta)k_2 p_{i-1}^* - k_2 p^*_i$$

and for the inactive polymers:

$$\frac{dp_1}{dt} = \delta W_{in}$$

$$\frac{dp_2}{dt} = \delta k_2 p_1^* \qquad (5.13)$$
$$\ldots$$
$$\frac{dp_i}{dt} = \delta k_2 p_{i-1}^*$$

In these equations the rate of initiation, W_{in}, is given by:

$$W_{in} = k_1(1-\xi) + k_1^* h_0 \exp(-k_1^* t) \qquad (5.14)$$

The concentration of chain propagation intermediates becomes instantaneously stationary:

$$\frac{dp_i^*}{dt} \approx 0 \qquad (5.15)$$

thus

$$p_i^* = (1-\delta)^i \frac{W_{in}}{k_2} \qquad (5.16)$$

and

$$p^* = \sum_{i=1}^{\infty} p_i^* = \frac{1-\delta}{\delta} \frac{W_{in}}{k_2} \qquad (5.17)$$

Using these relationships, we obtain from Eq. (5.11)

$$\frac{d\xi}{dt} = \frac{1}{\delta} W_{in} \qquad (5.18)$$

and from Eq. (5.13)

$$\frac{dp_1}{dt} = \delta W_{in} = \delta^2 \frac{d\xi}{dt}$$

$$\frac{dp_2}{dt} = \delta(1-\delta)W_{in} = \delta^2(1-\delta)\frac{d\xi}{dt} \qquad (5.19)$$

$$\cdots$$

$$\frac{dp_i}{dt} = \delta(1-\delta)^{i-1} W_{in} = \delta^2(1-\delta)^{i-1}\frac{d\xi}{dt}$$

As $k_2 >> W_{in}$, the stationary concentration of "polyenes" ending in activating double bonds (P_i^*) is very low, and the concentration of polyenes formed during the process is practically identical with that of the inactive products. It may be seen from Eq. (5.19), that the polyenes are formed proportionally to the conversion, in a geometrical distribution:

$$p_i = \delta^2(1-\delta)^{i-1}\xi \qquad (5.20)$$

The average length of the polyenes is inversely proportional to the probability parameter:

$$\bar{n} = \sum_{i=1}^{\infty} ip_i \Big/ \sum_{i=1}^{\infty} p_i = 1/\delta \qquad (5.21)$$

On the basis of Eqs. (5.14), (5.18), and (5.21), the rate equation of the conversion can be written as follows:

$$\frac{d\xi}{dt} = \bar{n}[k_1(1-\xi) + k_1^* h_0 \exp(-k_1^* t)] \qquad (5.22)$$

The solution of this differential equation is:

$$\xi = 1 - \exp(-\bar{n}k_1 t) + \{k_1^*\bar{n}h_0/(k_1^* - \bar{n}k_1)\}\{\exp(-\bar{n}k_1 t) - \exp(-k_1^* t)\} \quad (5.23)$$

Since $\bar{n}k_1 \ll k_1^*$ and $\bar{n}h_0 \ll 1$, the asymptotes for small conversions ($\xi \ll 1$) are, at the beginning of the process:

$$\xi = W_0 t \quad (5.24)$$

and after the consumption of structural defects:

$$\xi = \xi_0 + W_{st} t \quad (5.25)$$

where

$$W_0 = \bar{n}(k_1 + k_1^* h_0)$$

$$W_{st} = \bar{n}k_1 \quad (5.26)$$

$$\xi_0 = \bar{n}h_0$$

The results of this derivation, as described in Eq. (5.26), give quantities which can be experimentally determined (see Figure 5.1) in terms of important characteristics of the degradation: the average polyene length, $\bar{n}$; the defect concentration, h_0; and the rate constants of initiation (at random, k_1, and at the defects, k_1^*).

The polyenes are formed (and at the beginning of the degradation also exist) in geometrical distribution as described in Eq. (5.20). (Typical spectra are shown in Figure 2.24; in the later phases of degradation, the distribution of polyenes may change because of the secondary reactions.) Thus, the average polyene length can easily be determined using linear $\log p_i$ vs i plots. As already mentioned, $\bar{n}$ is usually small. One possible interpretation of this is that the activated complex preceding hydrogen chloride elimination is not formed exclusively — as is generally believed — by participation of the H atom in the β-position:

$$\begin{array}{cc}
\text{H} & \text{H} \\
\alpha| & |\beta \\
-\text{C}-\text{C}- \\
| & | \\
\text{Cl}\cdots\text{H}
\end{array} \longrightarrow
\begin{array}{cc}
\text{H} & \text{H} \\
\alpha| & |\beta \\
-\text{C}=\text{C}- \\
\end{array} \quad (5.27)$$

but in competition with this (in a ratio depending on the structure of the polymer and on the degradation conditions) by participation of the H atom in the γ-position as well:

$$\begin{array}{c}\overset{\beta}{CH_2}\\ \diagdown\overset{\alpha}{C}\diagup\quad\diagdown\overset{\gamma}{C}\diagup\\ \diagup\overset{|}{}\diagdown\quad\diagup\overset{|}{}\diagdown\\ H\quad Cl\text{---}H\quad Cl\end{array}\quad\rightarrow\quad\begin{array}{c}\overset{\beta}{CH_2}\\ \diagdown\overset{\alpha}{C}\diagup\quad\diagdown\overset{\gamma}{C}\diagup\\ \diagup\overset{|}{}\diagdown\qquad\diagup\overset{|}{}\diagdown\\ H\qquad\qquad Cl\end{array}\qquad(5.28)$$

From this activated complex an unstable cyclopropane ring is formed in the polymer chain as an intermediate, and in the course of its decomposition both activated and inactive products may form, e.g.,

$$-CH=CH-\underset{\underset{Cl}{|}}{CH}-CH_2-\qquad(5.29)$$

$$(P*)$$

$$\begin{array}{c}CH_2\\ \diagup\quad\diagdown\\ -CH-\!\!-\!\!-\!\!-\!\!C-CH_2-\\ \overset{|}{}\qquad\overset{|}{Cl}\end{array}$$

$$-CH_2-CH=\underset{\underset{Cl}{|}}{C}-CH_2-\qquad(5.30)$$

$$(P)$$

The ratio of activating and inactive products arising from a single elimination step (the stoichiometry of the reaction) is determined by the ratio of rate constants of parallel reactions leading to transition complexes [reactions (5.27) and (5.28)] and those [reactions (5.29) or (5.30)] of consecutive reactions (which are presumably quick in comparison to the conversion rate). The stoichiometric coefficients $(1-\delta)$ and δ thus formed appear in reactions (5.8)-(5.10) which give an overall view of the process and are rate-determining in terms of the conversion. These coefficients determine the value of the average polyene length.

Another possibility of chain termination is the formation of a cyclohexadiene ring. This may occur if there are three adjacent C—C bonds in cisoid-cis-cisoid conformation at the end of the propagating chain. Then, because of the favorable steric position, instead of further unzipping, cyclization may occur immediately after the last elimination step, which implies chain termination:

$$(5.31)$$

The defect concentration, h_0, and the rate constants k_1 and k_1^* can be determined by Eq. (5.26) from dehydrochlorination data if we have $\bar{n}$ from spectrophotometric measurements.

Experiments carried out on model compounds showed that in the presence of structural defects occurring in PVC, dehydrochlorination takes place more rapidly than in a compound of "regular" structure. The rate-increasing effect of the defects of different type and position, and the activation energy of degradation proceeding in their presence, are, however, different. It may be expected, therefore, that the various defects (double bonds in different position, tertiary carbon atoms, groups containing oxygen, etc.) affect the initial degradation rate of the polymer to a different extent at various temperatures. Indeed, defect concentrations (h_0 values) determined by using Eq. (5.26) slightly increase with increasing temperature.

It should be noted that great efforts have been made to determine the amount of the various defects in PVC; the procedure described here is the only direct method which gives the kinetic equivalent of the total amount of defects. Application of the method is possible only in solution degradation where careful rinsing out of evolved HCl can be performed; otherwise, for example, in the case of powder samples, the accelerating effect of HCl covers up the acceleration caused by the structural defects. This method can also be applied in the thermal degradation of some other polymers, e.g., polychloroprene, chlorobutyl rubber and chlorinated ethylene-propylene copolymers.

Secondary Processes

In the primary processes of PVC degradation, polyenes are formed which are highly reactive and may participate in various secondary reactions. Some important secondary reactions include cyclization, cross-linking, formation of volatile aromatic compounds, and oxidation of polyenes. The interaction between HCl and polyenes is of particular importance because it leads to the acceleration of the elimination process.

A very important secondary reaction of polyenes is *cross-linking*. The formation of an insoluble gel has been the subject of many investigations because it can be easily studied. Gel data can supply information on secondary reactions without direct (spectrophotometric) investigation of the polyenes.

If thermal degradation is carried out on solid samples, a measurable increase in molecular weight can be detected. Figure 5.2 shows the number average molecular weight of powder samples degraded at 180°C in a stream of argon as determined by osmometry and GPC.

The intrinsic viscosity also increases in the course of the process (Figure 5.3). At a certain cross-link density, part of the polymer becomes insoluble and a gel

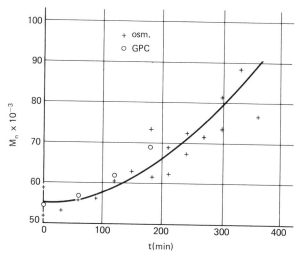

Figure 5.2. Change of molecular weight of PVC during thermal degradation (180°C, Ar). (Reprinted from reference 5.8, p. 350, by courtesy of Marcel Dekker, Inc.)

is formed. The amount of gel formed, γ, is indicated in Figure 5.3. Viscosity values after the gel point are those of the sol phase. The evaluation of the gel curve can be carried out by using the Charlesby-Pinner plot, Eq. (4.38) (see Figure 4.5, the line noted by Ar). However, when no scission occurs ($s = 0$), Eq. (4.38) can be written in another form:

$$2c = \frac{1}{\sigma + \sqrt{\sigma}} \tag{5.32}$$

or, introducing the designation

$$\alpha = \frac{2c}{P_{n,0}}$$

for the so-called cross-link density (each cross-link connects two monomeric units; α is the proportion of cross-linked units), we have from Eq. (5.32) for the post–gel point period:

$$\alpha P_{n,0} = \frac{1}{\gamma}\left(\frac{1}{\sqrt{1-\gamma}} - 1\right) \tag{5.33}$$

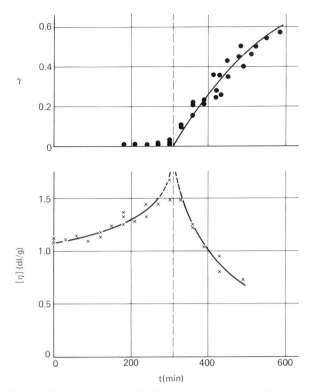

Figure 5.3. Change of intrinsic viscosity ($[\eta]$) and of the amount of gel (γ) formed during thermal degradation of PVC (180°C, Ar). (Reprinted from reference 5.8, p. 351, by courtesy of Marcel Dekker, Inc.)

Similarly, from Eq. (4.35) with $s = 0$ we have for the period before the gel point:

$$\alpha P_{n,0} = 2 \left(1 - \frac{P_{n,0}}{P_n} \right) \tag{5.34}$$

These equations enable us to plot both periods in a common diagram; Figure 5.4 shows the $\alpha P_{n,0}$ vs time representation of the molecular weight and gel data.

The appropriately transformed relationships fit surprisingly well with the experimental results of thermal cross-linking obtained by osmometry, GPC, and gel fraction measurements.   These show that the dependence of cross-link density on the degradation time is essentially linear.

There are two possible explanations of cross-linking during thermal degrada-

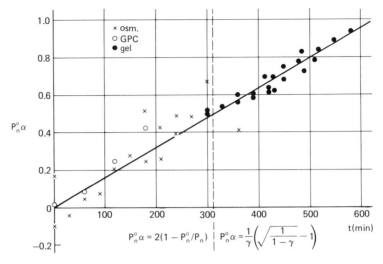

Figure 5.4. Dependence of cross-link density on degradation time (180°C, Ar). Dotted line represents the gel point (~310°C). (Reprinted from reference 5.8, p. 352, by courtesy of Marcel Dekker, Inc.)

tion of PVC: interchain HCl elimination, which would be a primary process, or intermolecular Diels-Alder reaction of the polyenes, a secondary process. One could distinguish between these mechanisms by adding maleic anhydride (MA) to swollen degraded PVC gels.

If cross-linking during thermal degradation occurs by Diels-Alder reaction, the addition of an excess of a strong dienophile (e.g., MA) to heated PVC gels should eliminate cross-links and bring the sample into solution. MA will preferentially react with the dienes in the degraded PVC chain and thus displace the Diels-Alder/*retro*-Diels-Alder equilibrium:

$$\text{(5.35)}$$

Experimentally, when thermal degradation of PVC is carried out for periods shorter than necessary for gel formation, treatment with maleic anhydride results in a decrease in molecular weight averages. Samples degraded beyond the gel point but only to relatively low gel content (up to 15% insoluble) become entirely soluble upon treatment with MA. Samples with higher gel content remain partly insoluble even after treatment with MA; however, their gel content decreases drastically (Figure 5.5).

These results indicate that a large portion of the cross-links formed during thermal degradation of PVC are reversible. Only Diels-Alder cross-linking of conjugated polyene sequences during heat degradation of PVC can explain partially the observed reversibility of cross-linking. The presence of other cross-linking mechanisms (e.g., recombination of polyenyl radicals, always present in degraded PVC) cannot be excluded because cross-linking has not been completely eliminated. However, even if all the cross-links were formed via the Diels-Alder reaction, full reversibility would not be expected because the polyenes involved in cross-links may undergo irreversible reactions as well.

Another important secondary process, *intramolecular isomerization* by *cyclization* of polyenes takes place under the conditions of thermal degradation. This process is reflected by the change in concentration of polyenes, which increases in proportion to hydrogen chloride elimination during the first period of degradation only (Figure 5.6). Later, the rate of concentration increase slows down, finally almost completely disappearing. The longer the polyene (the greater m is), the sooner this takes place. The effect of the process can also be observed in the ratio of polyene concentrations (Figure 5.7). With increasing conversion, the distribution shifts to shorter polyenes.

The kinetic treatment of the cyclization process must consider both the secondary decrease and the secondary formation of polyenes. It can be assumed

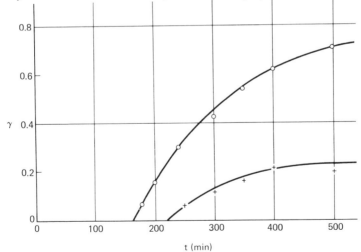

Figure 5.5. Effect of maleic anhydride (MA) treatment on the amount of gel fraction; o = without MA treatment; + = with MA treatment. (Reprinted, by permission, from reference 5.9.)

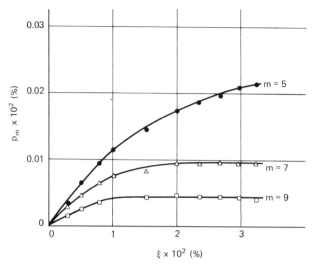

Figure 5.6. Relative amount of polyenes with five, seven, and nine conjugated double bonds plotted against dehydrochlorination conversion (210°C, Ar). (Reprinted from reference 5.8, p. 353, by courtesy of Marcel Dekker, Inc.)

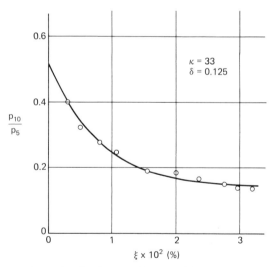

Figure 5.7. Concentration ratio of polyenes with ten and five double bonds as a function of dehydrochlorination conversion (210°C, Ar): δ corresponds to the initial average polyene length ($\bar{n} = 8$); κ is the rate constant ratio used for fitting differential equation (5.39) to the experimental points. (Reprinted from reference 5.8, p. 353, by courtesy of Marcel Dekker, Inc.)

that cyclization proceeds with equal probability on each $(m - 2)$ triad of double bonds in the polyene with m double bonds. The rate of decrease of polyene concentration is

$$W_m^- = -(m - 2)k_c p_m \tag{5.36}$$

where k_c is the cyclization rate constant. On the other hand, the rate of secondary formation of these polyenes is proportional to the concentration of all polyenes containing at least $(m + 3)$ double bonds:

$$W_m^+ = 2k_c \sum_{j = m + 3}^{\infty} p_j \tag{5.37}$$

For evaluation, the primary formation rate V_m^+ must also be taken into consideration. Its value is, according to Eq. (5.19),

$$V_m^+ = \delta^2 (1 - \delta)^{m - 1} \frac{d\xi}{dt} \tag{5.38}$$

The complete time dependence of p_m polyene concentration may therefore be written as

$$\frac{dp_m}{dt} = V_m^+ + W_m^+ + W_m^- \tag{5.39}$$

Differential equation (5.39) can be numerically integrated, using some approximations. As can be seen in Figures 5.6 and 5.7, the above description corresponds well to the experimental facts. The relative rate of cyclization, given by

$$\kappa \equiv k_c \bigg/ \frac{d\xi}{dt} \approx k_c / k_{HCl} \tag{5.40}$$

equals 33 for a PVC film sample at $210°C$ in an argon atmosphere.

The evolution of aromatic hydrocarbons, mainly benzene, in the thermal degradation of PVC has been observed by several authors. It seems evident that the formation of these aromatic compounds is related to the cyclization of polyenes. The time lag of the benzene formation curve (the induction period as compared to HCl evolution) is in accord with this reaction sequence (Figure 5.8).

For *benzene formation* without net chain scission, three possibilities exist: (1) the benzene is formed on the chain end; (2) the benzene is formed by chain

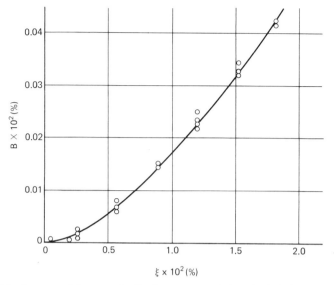

Figure 5.8. Amount of benzene (in % of monomer units) evolved during thermal degradation of PVC as a function of dehydrochlorination conversion (180°C, Ar). (Reprinted from reference 5.8, p. 356, by courtesy of Marcel Dekker, Inc.)

scission, but the free chain ends recombine; and (3) free chain ends are not formed at all in this process. As the polyene sequences are formed statistically on the polymer chain, the first possibility is expected to be insignificant. For the remaining two cases, the following mechanisms are considered:

$$(5.41)$$

and

$$(5.42)$$

Probably the most important secondary reaction of PVC degradation is the *interaction of the evolved HCl with the polyenes.* HCl catalysis is one of the most debated questions of PVC degradation. The mechanism of catalysis has not yet been clarified despite the large number of papers on the subject; even the experimental results seem to be controversial. In solution degradation, no accelerating effect can be observed at all (see Figure 5.1). In the degradation of film or powder samples, however, the rate increases with increasing thickness. As already mentioned, this phenomenon is caused by the incomplete removal of the evolved HCl.

HCl affects both the rate of HCl loss and the distribution of polyene sequences. The participation of polyenes in HCl catalysis was suggested a decade ago, and it has been found that the rate of HCl-catalyzed degradation increases not only with increasing HCl pressure but also with increasing polyene content. Fast proton exchange of the polyenes was demonstrated using tritium-labeled HCl. Recently, proton exchange and conversion of shorter polyene sequences to longer ones were observed during treatment of chemically degraded PVC with trifluoro-acetic acid.

In the special device (using evacuated and sealed h-shaped ampules which allowed the study of degradation with and without freezing out of the evolved HCl by application or removal of a liquid air cooler), experiments have been carried out to clarify the role of proton exchange in HCl catalysis. Results of dehydrochlorination experiments in vacuo with freezing out of HCl are identical with those which can be obtained in argon flow (Figure 5.9). However, the removal of liquid air (i.e., evaporating the HCl which then contacts the sample) results in a dramatic acceleration of dehydrochlorination.

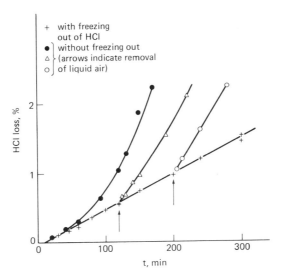

Figure 5.9. Dehydrochlorination conversion curves at 180°C, obtained in experiments in the presence or absence of HCl. (Reprinted, by permission, from reference 5.14.)

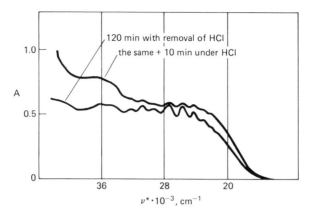

Figure 5.10. Spectra of PVC samples degraded in evacuated ampules with and without freezing out of the evolved HCl (180°C). (Reprinted, by permission, from reference 5.14.)

According to spectrophotometric measurements, when HCl is removed polyene distribution shifts slowly toward shorter sequences because of secondary reactions. In the presence of HCl, the shapes of the spectra are markedly dissimilar and the resolution of the peaks corresponding to the individual polyenes diminishes. This is especially pronounced when the HCl is allowed to react with the polyenes of predegraded PVC (Figure 5.10).

To facilitate comparison of the spectra taken under different degradation conditions, the ratios of absorbances at $\nu^* = 20,000$ and $28,000$ cm^{-1} (A_{20}/A_{28}) and at $\nu^* = 36,000$ and $28,000$ cm^{-1} (A_{36}/A_{28}) are plotted against degradation

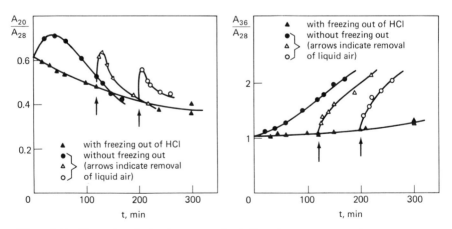

Figure 5.11. The ratio of absorbances at the indicated wave numbers for PVC samples degraded at 180°C. (Reprinted, by permission, from reference 5.14.)

time (Figure 5.11). A_{20} corresponds to 12–13, A_{28} to 5–6, and A_{36} to 3 conjugated double bonds in the polyene; thus A_{20}/A_{28} is a measure of the concentration of long polyenes, and A_{36}/A_{28} the concentration of short polyenes, relative to medium ones.

During degradation with the removal of HCl, the ratio A_{20}/A_{28} slowly decreases due to polyene consuming secondary reactions. In the presence of HCl, however, a sharp increase in A_{20}/A_{28} is observed indicating the formation of longer polyenes.

In explaining the experimental results, we must take into account that the primary reactions of thermal degradation of PVC involve relatively slow initiation, allyl-activated zip-elimination of HCl, and termination of unzipping. In the presence of HCl and other acids, proton exchange may lead to the migration of the polyene, and a chlorine atom becomes allyl activated. Reinitiation of allyl-activated unzipping occurs as shown in Figure 5.12. The newly formed double bonds are conjugated with the original polyene, thus the average polyene length increases.

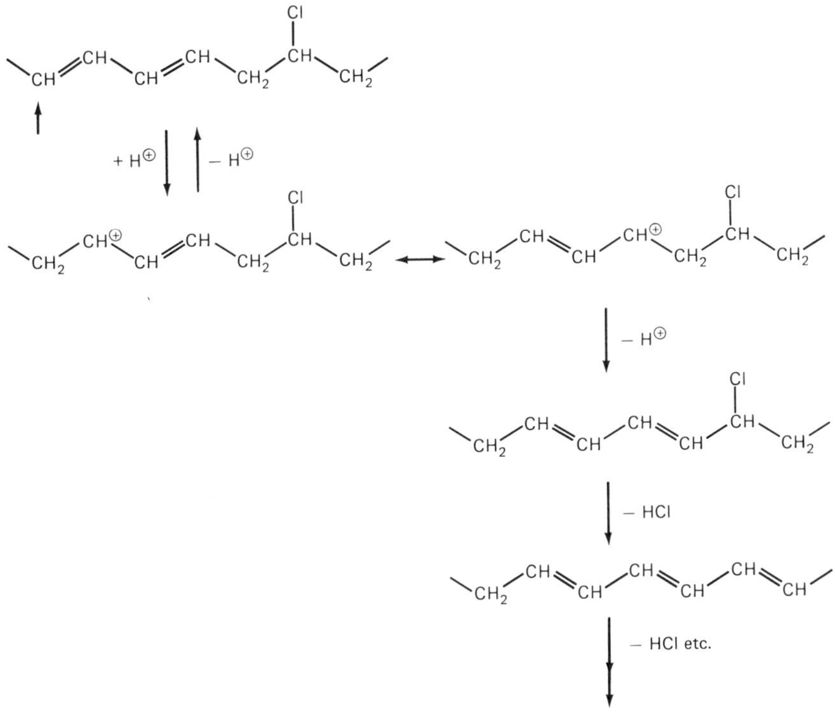

Figure 5.12. Reinitiation scheme of HCl catalysis in polyvinyl chloride degradation. (Reprinted, by permission, from reference 5.14.)

Reinitiation of earlier terminated polyenes is not the sole effect of HCl. As shown in Figure 5.11, after the sudden increase the A_{20}/A_{28} ratio decreases faster in the presence of HCl than in vacuo, and at the same time a rapid increase in the A_{36}/A_{28} ratio occurs. These observations indicate that the polyene consuming secondary reactions are accelerated by HCl.

The reinitiation mechanism and the simultaneous acceleration of secondary reactions by HCl clarify the apparent controversy about the increase or decrease of the average polyene length in the presence of HCl. In some cases polyene growth by reinitiation predominated, while under other conditions the effect of HCl-catalyzed secondary reactions was more pronounced.

In addition to the above two effects, HCl may influence PVC degradation in other ways, e.g., by catalyzing primary HCl loss.

5.3 THERMO-OXIDATIVE DEGRADATION OF PVC

The rate of evolution of HCl is considerably higher in the presence of oxygen than in an inert atmosphere or in a vacuum. Thermo-oxidative degradation of solid PVC samples is an accelerating process. Acceleration is especially pronounced at high temperatures (above 180°C) and high extents of HCl loss. A characteristic difference from thermal degradation may be discerned in the color and UV-visible spectra of samples degraded in inert atmospheres and those degraded in oxygen. At identical extents of HCl loss, PVC degraded in the presence of oxygen has a lighter color. Absorption maxima corresponding to polyenes of different lengths, clearly visible in the spectrum of thermally degraded PVC, cannot be observed in the case of oxidative degradation which gives unresolved patterns. A further difference is that in the course of thermal degradation practically no scission of the main chain occurs, whereas in the presence of oxygen, chain scissions play an important role (see Figure 4.5; the intercept of the Charlesby-Pinner plot is proof positive for chain scission).

For a better understanding of the thermo-oxidative degradation of PVC, knowledge of the kinetics and mechanism of *polyene oxidation* is required. This is a rather complex process, but fortunately it can be studied separately from the even more complicated process of thermo-oxidative degradation.

The consumption of the individual polyenes can be followed by UV-visible spectrophotometry (Figure 2.24). Based on the changes of spectra, surprisingly simple oxidation kinetics were observed when experiments were carried out in pure oxygen at atmospheric pressure. In such cases, oxidation obeys a first-order rate law, i.e., the log $[(A - A_\infty)/(A_0 - A_\infty)]$ vs time plot for a given polyene length is linear. The slope of such plots, i.e., the rate constant for the consumption of polyene sequences, turned out to be proportional to the polyene length. This is illustrated in Figure 5.13 where the product of polyene

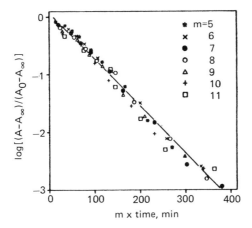

Figure 5.13. Logarithmic plot of polyene consumption as a function of polyene length and oxidation time (predegradation at 210°C, oxidation at 60°C). (Reprinted, by permission of Applied Science Publishers Ltd., from reference 5.24.)

length and oxidation time ($m \times t$) is used as the independent variable. All points are seen to lie on the same straight line. Thus, the reaction rate constant can be characterized by a single value, namely, the slope W of this straight line, which may be considered the overall oxidation rate. When the experimental conditions are changed, behavior typical of radical oxidation processes is observed. Thus, in the presence of a radical initiator, e.g., azobisisobutyronitrile, W is proportional to the square root of the initiator concentration, indicating bimolecular termination. W is practically independent of partial oxygen pressure above 0.5 atm due to the fast conversion of alkyl (i.e., polyenyl) radicals into peroxy radicals

$$\tag{5.43}$$

which participate predominantly in chain termination.

In the oxidation of polyenes, a distinction should be made between intermolecular and intramolecular chain propagation steps. In the former, oxidative chain propagation proceeds from one polyene to another, whereas in the latter the peroxy radical reacts with a nearby double bond in the same molecule, forming cyclic peroxides – most probably six- and five-membered rings:

$$(5.44)$$

With the usual approximations applied for radical chain reactions, a kinetic equation for the *initiated oxidation of polyene* sequences has been deduced:

$$\log (p_{m0}/p_m) = Wmt$$

$$W = (2k_1 f)^{1/2} I^{1/2} [k_2/(2k_4)^{1/2}] \qquad (5.45)$$

where W is the overall rate of polyene consumption; p_{m0} and p_m are the concentrations of m-long polyenes at the beginning of oxidation and after time t, respectively; k_1 is the rate constant for the decomposition of the initiator; f is the initiation efficiency; k_2 is the rate constant for intermolecular propagation related to one double bond in the polyene sequence; k_4 is the rate constant for termination; and I is the initiator concentration.

The *autoxidation of polyene* sequences in solution is similar to that of initiated oxidation, but the rate of initiation changes with oxidation time due to changes in the concentration of the initiating species; in other words, the change results from degenerate chain branching. Fortunately, proportionality between consumption rate and polyene length exists to a good approximation in this case. At the beginning of the oxidation the rate increases with increasing oxidation time, but at later stages it decreases approximately according to the first-order rate law.

The amount of hydroperoxides as well as the sum of hydroperoxide and dialkyl peroxide concentrations can be measured selectively. In the course of autoxidation of polyene sequences, the total amount of peroxides increases gradually, whereas the concentration of hydroperoxides reaches a maximum shortly after the onset of oxidation and then decreases very slowly, remaining almost constant for a long period. The amount of dialkyl-type, relatively stable, cyclic peroxides is considerably higher than the concentration of hydroperoxides. At the end of

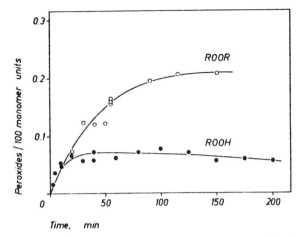

Figure 5.14. Formation of hydroperoxides and dialkyl (cyclo) peroxides during autoxidation of polyenes in degraded PVC (predegradation at 210°C, oxidation at 100°C). (Reprinted, by permission of Applied Science Publishers Ltd., from reference 5.24.)

oxidation, the total amount of peroxides exceeds one-quarter of the double bonds initially present in the reaction (Figure 5.14).

The presence of polyenes in the polymer accelerates the dehydrochlorination in the *thermo-oxidative degradation* of PVC, as shown in Figure 5.15. The dehydrochlorination is faster in O_2 than in Ar even when the sample has not been predegraded (0 min). The dehydrochlorination rate in O_2, however, increases with increasing predegradation times (30 and 180 min), i.e., with increasing polyene content. Thus polyene oxidation has a "feedback" effect on the primary processes.

Besides HCl evolution other changes such as oxygen absorption and weight decrease can also be followed during PVC oxidation, as shown in Figure 5.16. If the total amount of oxygen absorbed were bound to the polymer, this would largely compensate for the weight loss caused by loss of HCl. The experiments in pure oxygen, however, show higher weight losses than expected. This indicates that in addition to HCl, other volatile products are formed, some of which also contain oxygen (e.g., water). The amount of these products can be calculated from the weight of absorbed oxygen, eliminated HCl, and the measured weight.

The UV-visible spectrum of PVC degraded under thermo-oxidative conditions (Figure 5.17) is basically different from the spectrum of thermally degraded samples but similar to the spectrum of oxidized polyenes (Figure 2.24).

The thermo-oxidative degradation of PVC is a complicated chemical process even in the initial stages (up to a few percent HCl loss). The mechanism is not

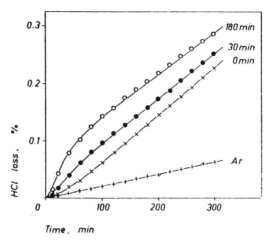

Figure 5.15. Loss of HCl during oxidation of virgin (0 min) and predegraded (180°C in Ar for 30 min and 180 min) PVC powder samples; as a comparison, HCl loss under neutral atmosphere is also given (Ar). (Reprinted, by permission of Applied Science Publishers Ltd., from reference 5.24.)

known in detail, but the main reaction routes which involve consecutive, competitive, and "feedback" steps are clear. At the temperatures usually applied (above 160°C), the main reaction *routes are as follows*:

1. Primary HCl loss and polyene formation take place in a similar process in oxygen and in inert atmospheres.
2. The polyene sequences are readily oxidized in radical chain reactions.
3. The peroxides formed decompose rapidly.
4. The radicals formed in the course of oxidation attack intact monomer units and *initiate further* HCl loss.

These basic steps are represented in Figure 5.18 which is a very simplified (minimum) scheme of thermo-oxidative PVC degradation.

As shown in Figure 5.19 for a *high conversion* measurement, thermo-oxidative HCl loss is autoaccelerating; thus its kinetics can be described by the following formal autocatalytic equation:

$$d\xi/dt = (k_s + k\xi)(1 - \xi/\xi_\infty)$$ (5.46)

where ξ is the conversion of HCl loss; t is the time; k_s is the steady-state rate (extrapolated to the beginning of the degradation); and k and ξ_∞ are formal constants. High conversion experiments are usually carried out above 200°C,

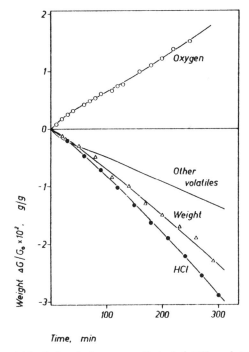

Figure 5.16. The weight of absorbed oxygen, eliminated HCl, and polymeric residue, as well as the calculated amount of volatiles (except HCl) during thermo-oxidative degradation of PVC (180°C, 1 atm O_2). (Reprinted, by permission of Applied Science Publishers Ltd., from reference 5.24.)

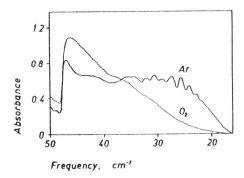

Figure 5.17. Visible spectra of degraded PVC. Degradation of powder samples at 200°C to 0.6% HCl loss in 1 atm Ar and O_2. Photometry in tetrahydrofuran solution (9 g/liter). (Reprinted, by permission of Applied Science Publishers Ltd., from reference 5.24.)

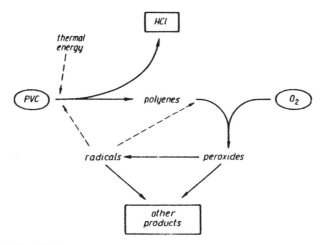

Figure 5.18. Simplified reaction scheme for thermo-oxidative PVC degradation. Dashed arrows indicate initiation. (Reprinted, by permission of Applied Science Publishers Ltd., from reference 5.24.)

where the transition period is negligibly short, so the initial rate can be taken as the steady-state rate. Appreciable deviation is observed only at conversions above 70% HCl loss if $\xi_\infty = 1$ is used.

5.4 DEGRADATION OF PVC UNDER DYNAMIC CONDITIONS

PVC can be processed only in the presence of stabilizers because of its relatively low decomposition temperatures. During processing, not only heat but shear as well affects the polymer. The performance of PVC under dynamic conditions is of great importance. Although measuring equipment exists which is suitable for testing PVC under dynamic conditions, the first aim of these instruments (which include torque rheometers, e.g., Haake Rheocord and Brabender Plastograph, and extrusiometers) is to determine the fusion, rheological, and processing characteristics of PVC compounds. The so-called *Brabender Stability Test* determines the end point of effective stabilization and the beginning of severe cross-linking but provides no insight into the processes which cause these phenomena.

The Brabender test is not standardized; thus there are some deviations in its application in various (mainly industrial) laboratories. A typical test is the following: In a 50 cm³ capacity mixing chamber with roller-type blades, 60 g of PVC is melted and mixed at 190°C at 50 rpm. After the fusion point is reached, the torque decreases to a value characteristic of the undegraded polymer. Later, due to cross-linking of the degraded material, the torque increases (see curve *M* in Figure 5.20). The time at which cross-linking begins, determined by the

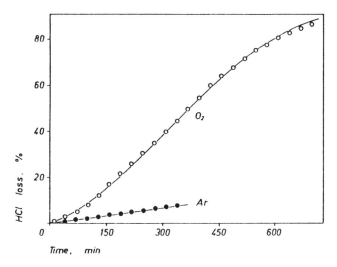

Figure 5.19. Loss of HCl from PVC powder degraded to high conversions at 200°C in 1 atm O_2 (for comparison, degradation in Ar is also given). (Reprinted, by permission of Applied Science Publishers Ltd., from reference 5.24.)

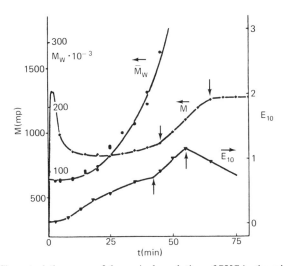

Figure 5.20. Characteristic curves of dynamic degradation of PVC in the mixing chamber of a torque rheometer at 190°C: M, torque; $\overline{M}_w$, molecular weight; and E_{10} extinction of polyenes with $n = 10$. The first arrow on the M curve ($\sim$45 min) corresponds to the so-called Brabender Stability Time; the second arrow on the E_{10} curve ($\sim$60 min), to the gel point.

intersection of the horizontal and increasing parts of the diagram, is used to characterize the effectiveness of the added stabilizer (in Figure 5.20, the first arrow on curve M, at about 45 min).

Dynamic degradation is usually carried out in air. Infrared measurements of the treated polymer show a very low carbonyl content which may indicate that oxygen is present only in low concentration in the melt and does not play an essential role in the dynamic test.

In Figure 5.20, the change of E_{10}, the extinction of polyenes with ten conjugated double bonds, is shown as a function of treatment time. On this curve, three stages can be distinguished: in the first stage, extinction increases moderately; then the rate of polyene formation, i.e., the rate of degradation, accelerates suddenly (first arrow on the E_{10} curve); finally, when the gel point is reached (second arrow; see also the molecular weight curve) and the material becomes partially insoluble, the extinction of the sol phase decreases.

The weight average molecular weight ($\overline{M}_w$) does not change at the beginning of the test, but later increases rapidly and approaches infinity at the gel point. The Charlesby-Pinner plot of the gel data (see Figure 4.5) shows that in addition to cross-linking, a chain scission process occurs simultaneously.

The torque vs time curve has a well-known shape. The point at which it shows a relatively sharp break (it starts to increase rapidly due to cross-linking) lies at a longer degradation time than the break on the extinction curve. This indicates that the sudden increase in the rate of polyene formation cannot be caused by

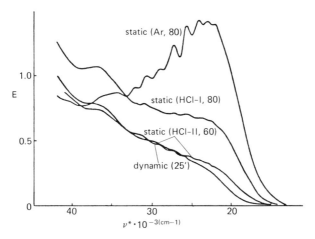

Figure 5.21. UV spectra of a sample from the dynamic test and of samples degraded under static conditions, under Ar atmosphere (Ar), and in sealed evacuated tubes (HCl pressure in the tube at the end of measurement was 0.6 atm for HCl-I and 1.9 atm for HCl-II). The numbers in parentheses are degradation times; for better comparison, they are about half of the corresponding gel points.

the torque increase; consequently, we must look for its origin in the chemical degradation process.

In the dynamic test, HCl is not effectively removed. One can assume that the rate increase may be related to the catalytic effect of HCl. This possibility can be confirmed by comparison of the UV spectra of samples taken from the dynamic test and samples degraded in an argon atmosphere or in the presence of HCl (Figure 5.21; because of differences in rates, the spectra are compared at times equivalent to half of the individual gel points). The different HCl pressures in experiments I and II were achieved by using different amount of PVC in tubes having the same volumes. It can be established that with increasing HCl pressure, the UV spectra of samples from the static measurements approach that of the sample treated in the mixing chamber. This shows that HCl is dissolved in the melt and plays an important role in the degradation process taking place in the mixing chamber.

REFERENCES

5.1. Abbås, K. P. and Sörvik, E. M. On the thermal degradation of poly(vinyl chloride). I. An apparatus for investigation of thermal degradation. *J. Appl. Polymer Sci.* 17:3567 (1973).

5.2. Abbås, K. P. Characterization of polyene sequences in poly(vinyl chloride). *J. Macromol. Sci.–Chem.* A12:479 (1978).

5.3. Baum, P. and Wartman, L. H. Structure and mechanism of dehydrochlorination of polyvinyl chloride. *J. Polymer Sci.* 28: 537 (1958).

5.4. Braun, D. Recent progress in the thermal and photochemical degradation of poly-(vinyl chloride). In (Geuskens, G., ed.) *Degradation and Stabilization of Polymers.* London: Applied Science Publishers, 1975.

5.5. Geddes, W. C. Mechanism of PVC degradation. *Rubber Chem. Tech.* 40:177 (1967).

5.6. Guyot, A. and Bert, M. Sur la degradation thermique du polychlorure de vinyle. VI. Etapes initiales–etude generale preliminaire. *J. Appl. Polymer Sci.* 17:753 (1973).

5.7. Hjertberg, T. and Sörvik, E. M. On the influence of HCl on the thermal degradation of poly(vinyl chloride). *J. Appl. Polymer Sci.* 22:2415 (1978).

5.8. Kelen, T. Secondary processes of thermal degradation of PVC. *J. Macromol. Sci.– Chem.* A12:349 (1978).

5.9. Kelen, T., Iván, B., Nagy, T. T., Turcsányi, B. and Tüdös, F. Reversible crosslinking during thermal degradation of PVC. *Polymer Bulletin* 1:79 (1978).

5.10. Kolinsky, M. Effect of vinyl chloride polymerization conditions on polymer thermal stability. *J. Macromol. Sci.–Chem.* A11:1411 (1977).

5.11. Lévai, G., Ocskay, G. and Szebeni, S. Investigation of thermal degradation of PVC in the solid state. *J. Macromol. Sci.–Chem.* A12:467 (1978).

5.12. Mayer, Z. Thermal decomposition of poly(vinyl chloride) and of its low-molecular-weight model compounds. *J. Macromol. Sci.–Revs. Macromol. Chem.* C10:263 (1974).

5.13. Minsker, K. S., Lisitskii, V. V., Kronman, A. G., Gataullin, R. F. and Chekushina, M. A. Conversions of structural groups during the degradation of vinylchloride and vinyl-acetate polymers. *Polymer Science USSR* 22:1228 (1980).

5.14. Nagy, T. T., Kelen, T., Turcsányi, B. and Tüdős, F. The reinitiation mechanism of HCl catalysis in PVC degradation. *Polymer Bulletin* 2:77 (1980).

5.15. Nagy, T. T., Iván, B., Turcsányi, B., Kelen, T. and Tüdős, F. Crosslinking, scission and benzene formation during PVC degradation under various conditions. *Polymer Bulletin* 3:613 (1980).

5.16. Nass, L. I. Theory of degradation and stabilization mechanisms. In (Nass, L. I., ed.) *Encyclopedia of PVC*, Vol. 1, Ch. 8. New York: Marcel Dekker, 1976.

5.17. Ocskay, G., Nyitrai, Zs., Várfalvi, F. and Wein, T. Investigation of degradation processes in PVC based on the concomitant colour changes. *European Polymer J.* 7:1135 (1971).

5.18. Onozuka, M. and Asahina, M. On the dehydrochlorination and the stabilization of polyvinyl chloride. *J. Macromol. Sci.–Revs. Macromol. Chem.* C3:235 (1969).

5.19. Pukánszky, B., Nagy, T. T., Kelen, T. and Tüdős, F. Comparison of dynamic and static degradation of poly(vinyl chloride). *Preprints of Third International Symposium on Polyvinylchloride.* Cleveland, 1980.

5.20. Starnes, W. H., Jr. Mechanistic aspects of the degradation and stabilization of poly(vinyl chloride). In (Grassie, N., ed.) *Developments in Polymer Degradation–3.* London: Applied Science Publishers, 1981.

5.21. Troitskii, P. P., Troitskaya, L. S., Denisova, V. N. and Lusinova, Z. P. Kinetics of initial stage of the thermal dehydrochlorination of poly(vinyl chloride). *Polymer Journal* 10:377 (1978).

5.22. Tüdős, F. and Kelen, T. Investigation of the kinetics and mechanism of PVC degradation. In (Saarela, K., ed.) *Macromolecular Chemistry–8.* London: Butterworths, 1973.

5.23. Tüdős, F., Kelen, T., Nagy, T. T. and Turcsányi, B. Polymer-analogous reactions of polyenes in poly(vinyl chloride). *Pure and Applied Chemistry* 38:201 (1974).

5.24. Tüdős, F., Kelen, T. and Nagy, T. T. Thermo-oxidative degradation of poly(vinyl chloride). In (Grassie, N., ed.) *Developments in Polymer Degradation–2.* London: Applied Science Publishers, 1979.

Chapter 6
Oxidative Degradation

Oxidation of hydrocarbons is an important process which has long been known. The reaction of organic compounds initiated by the attack of molecular oxygen is called *autoxidation*. Oxidation is important not only because it is used in the petrochemical industry for the production of useful oxygenated chemicals and raw materials, but also because it is often responsible for the deterioration of polymers.

Autoxidation of unsaturated compounds such as oils, rubber, etc., had been recognized and the main characteristics of the process [e.g., the autoacceleration (sigmoidal) type of oxygen absorption vs time curve, formation of peroxides, etc.] had already been observed in the last century. Intensive studies of rubber oxidation, pioneered by workers of the British Rubber Producers Association in the early 1940s (Bolland and Gee, Bateman, and others), led to the development of a *basic autoxidation scheme* (BAS). This scheme included all the important steps in oxidation (it consists of a radical chain reaction with degenerate branching) and could explain many features of the process. Recent theories of polymer oxidation are based on BAS, with minor modification and extensions.

As a result of the low rate of radical generation at ambient conditions, the rate of polymer oxidation is generally quite low. Irradiation or other factors which increase radical production will, of course, increase the rate of polymer oxidation. Peroxides or other oxygenated compounds incorporated into a polymer during its synthesis or processing may increase its susceptibility to oxidation.

Polymer oxidation reactions generally show an induction period during which no visible changes occur in the material. However, after the induction period, the rate of oxidation increases rapidly (autocatalysis): the first formed oxidation products accelerate the further degradation.

The oxidation of polymers causes a deterioration of physical properties.

Often, oxidation leads to decreased molecular weight and discoloration of the polymer. The oxidized polymers have lower mechanical strength. Oxidation also changes the electrical properties of polymers. In general, oxidized polymers have higher conductivity and a higher dielectric constant than unoxidized materials. When the oxygen concentration is small, oxidation can cross-link the polymer. Such materials are invariably stiffer, more brittle, and less prone to creep than the parent materials.

6.1 GENERAL REMARKS

The oxidative degradation of polymers is a free radical chain reaction. An important difference from some other chain reactions is that besides the usual three steps, i.e., initiation, propagation, and termination, two additional important steps must be considered here: conversion of the formed hydrocarbon radicals to peroxy radicals (this is the main oxygen consuming reaction) and degenerate chain branching (this is the reaction responsible for the autoacceleration character of the process). A scheme including the most important steps in polymer oxidation is shown in Figure 6.1. It is to be noted that this is not a complete oxidation scheme; there are several other reactions which may take

I. *Initiation*:

$$RH + O_2 \longrightarrow R^{\cdot} + HO_2^{\cdot} \qquad \text{(I.1)}$$
$$\text{Initiator} \longrightarrow 2f'R^{\cdot\cdot} \qquad \text{(I.2)}$$
$$R^{\cdot\cdot} + RH \longrightarrow R^{\cdot} + R'H \qquad \text{(I.3)}$$
$$HO_2^{\cdot} + RH \longrightarrow R^{\cdot} + H_2O_2 \qquad \text{(I.4)}$$

II. *Radical conversion (stabilization)*:

$$R^{\cdot} + O_2 \longrightarrow RO_2^{\cdot} \qquad \text{(II)}$$

III. *Chain propagation*:

$$RO_2^{\cdot} + RH \longrightarrow R^{\cdot} + RO_2H \qquad \text{(III.1)}$$
$$RO_2^{\cdot} + RH \longrightarrow R^{\cdot} + \text{products} \qquad \text{(III.2)}$$

IV. *Degenerate chain branching*:

$$RO_2H \longrightarrow f(RO^{\cdot} + HO^{\cdot}) + (1-f)(RO + H_2O) \qquad \text{(IV.1)}$$
$$RO^{\cdot} + RH \longrightarrow R^{\cdot} + ROH \qquad \text{(IV.2)}$$
$$HO^{\cdot} + RH \longrightarrow R^{\cdot} + H_2O \qquad \text{(IV.3)}$$

V. *Termination*:

$$R^{\cdot}, RO_2^{\cdot}, \text{etc.} \longrightarrow \text{products} \qquad \text{(V)}$$

Figure 6.1. Simplified general scheme of polymer oxidation.

place during the oxidation of a polymer, depending on the chemical structure of the investigated polymer, the type and amount of contaminants and additives present in the system, etc. Details which would explain some of the important features such as the formation of volatile products, the proportions of various oxygen containing groups in the polymers, etc., are not included in this generalized and simplified scheme.

I. *Initiation* by direct reaction of the substrate and molecular oxygen, as shown in (I.1) of Figure 6.1, is for most polymers thermodynamically and kinetically unfavorable. Reaction (I.1) is highly endothermic and must be very slow at low temperatures. If the oxidation had no other source of radicals than (I.1), no acceleration would occur and the process would be harmless and unimportant. However, the degenerate chain branching reaction (IV.1), i.e., the decomposition of the hydroperoxides into radicals, takes over the role of initiation in the later phase of oxidation. (In a radical chain reaction, chain branching is a reaction which produces more than one radical from one radical; degenerate chain branching produces a radical or radicals from a nonradical intermediate of the process.) When peroxides are present in the original polymer, e.g., those which are formed during processing or storage, reaction (IV.1) may dominate as initiation, even in the initial phase of oxidation.

The probability of reaction (I.1) occurring is higher when the polymer contains reactive hydrogens. For example, the lower oxidative stability of polypropylene compared to that of polyethylene can be explained by the tertiary H atoms present only in polypropylene. Figure 6.2 shows carbonyl absorption vs time curves for the oxidation of deuterated polypropylene samples. As can be seen, the sample containing deuterium atoms in place of the tertiary hydrogen has a much longer induction period and a slower rate than the other samples. This isotope effect shows, however, that the lower dissociation energy of the tertiary $C-H$ bond plays a role not only in the (I.1) initiation step (induction period) but also in the (III.1) and (III.2) propagation steps (rate) of the oxidation.

In fundamental studies of the oxidative degradation of polymers, the rate of initiation must be constant and measurable. The study of polymer autoxidation is complicated by the lack of a precise measure of the rate of initiation. In accelerated degradation studies, free radical initiators are used to initiate free radical chain reactions. Such initiators are well known, and a wide variety of azo compounds and peroxide compounds which thermally decompose to give free radicals are commercially available. The use of such compounds as initiators of chain reactions simplifies the study of degradation because the rate of radical production is constant and can be readily measured. In Figure 6.1, (I.2) represents the initiator decomposition reaction. Here f' denotes the efficiency of initiation (radical yield): after dissociation of the initiator molecule the paired radicals may either recombine in the "cage" or separate to give free radicals.

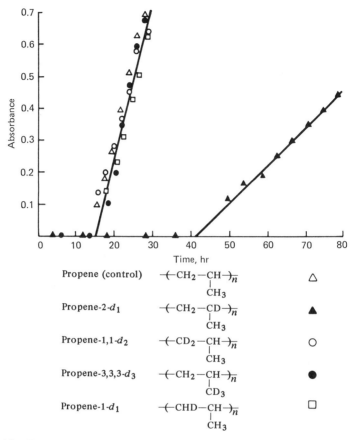

Figure 6.2. Changes of the carbonyl absorption during thermal oxidation (100°C) of polypropylene samples with deuterium atoms in different positions. (Reprinted from reference 6.20, p. 482, by courtesy of Marcel Dekker, Inc.)

The cage effect is an important consideration, especially when studying polymer oxidation in viscous media. It has long been known that the cage effect becomes more important as the viscosity of the system increases. In a viscous medium, radicals produced in close proximity have little chance to separate; hence, radical recombination (termination) can occur. Thus, in viscous systems, the probability of radical production is decreased; this is so not only in reaction (I.2) but also in (IV.1) which is, in a narrower sense, an initiation reaction as well.

The very mobile products of the initiation steps easily react with the (active) hydrogens of the polymer. Thus, the $HO_2 \cdot$ radical formed in (I.1) transforms to H_2O_2 in reaction (I.4), the $R' \cdot$ fragment of initiator formed in (I.2) turns into

R′H, usually a volatile product, in reaction (I.3).   Both reactions lead to the formation of the R· hydrocarbon (alkyl) radical.

II. *The conversion of the hydrocarbon radical to a peroxy radical* is an important step in the oxidation; namely, it is the step in which the majority of oxygen is absorbed by the polymer.   As soon as an alkyl radical is generated in an initiation reaction, it nearly always reacts readily with oxygen.   Oxygen itself is a diradical and the reaction of an alkyl radical with oxygen, reaction (II) in Figure 6.1, is essentially a radical coupling reaction.   Oxygen is 10^6 times as reactive as styrene toward the polystyrenyl radical, even though styrene itself is very reactive toward free radicals.

The saturated alkylperoxy radicals are quite stable below 300-400°C; thus reaction (II) is sometimes called *radical stabilization*.   The rate of this reaction, however, depends on the concentration of the oxygen "inside" the polymer, i.e., on its pressure "outside" and on the ease of diffusion.   Thus, with low oxygen pressure and/or with high sample thickness, the oxidation may become diffusion controlled.   As the life of the alkyl radical is lengthened with low oxygen concentration, depolymerization, chain transfer, and cross-linking may become more probable.

III. *Chain propagation* in polymer oxidation consists of the hydrogen abstraction reaction of the peroxy radicals.   With increasing rate constant of this reaction, the rate of oxidation increases and the kinetic chain length becomes greater. We included two versions of this reaction in our scheme in Figure 6.1: (III.1) describes the reaction with hydroperoxide formation, and (III.2) without hydroperoxide formation.   The ratio of the rate constants of these reactions (together with the f cage factor of the hydroperoxide decomposition) determines the degree of acceleration of the oxidation: the higher k_{31}/k_{32} and f, the more the process accelerates.

Reactions leading to conversion of various radicals to an alkyl radical – (I.3) for the initiator radical, (I.4) for HO_2 ·, (IV.2) for the alkoxy, and (IV.3) for the hydroxy radicals – are essentially the same type of hydrogen abstraction reactions as the propagation steps (III.1) and (III.2).   The reactivity of hydrogen toward free radicals increases in the order primary < secondary < tertiary, in accordance with the decrease in bond dissociation energies (see Figure 6.2).   Because of the high reactivity of the tertiary hydrogens, intramolecular propagation may be a major reaction path in the oxidation of some polymers such as polypropylene and polystyrene.   The examination of the structure of polypropylene hydroperoxide has shown that many of the hydroperoxide groups are intramolecularly hydrogen bonded.   The formation of hydroperoxide sequences is possible, however, only in cases which are sterically favored:

$$\underset{\overset{|}{OO\cdots\cdots H}}{\overset{\overset{CH_3}{|}}{\sim\!-\!C\!-\!CH_2}}\!-\!\underset{\overset{|}{H}}{\overset{\overset{CH_3}{|}}{C}}\!-\!CH_2\!-\!\underset{}{\overset{\overset{CH_3}{|}}{C}}\!\sim\!\overset{O_2}{\rightarrow}\;\underset{\overset{|}{OOH}}{\overset{\overset{CH_3}{|}}{\sim\!-\!C\!-\!CH_2}}\!-\!\underset{\overset{|}{OO\cdots\cdots H}}{\overset{\overset{CH_3}{|}}{C}}\!-\!CH_2\!-\!\underset{}{\overset{\overset{CH_3}{|}}{C}}\!\sim\!\overset{O_2}{\rightarrow}\quad(6.1)$$

Intramolecular peroxy radical attack is highly efficient at the β-position, somewhat less at the γ-position, and of little or no significance in other positions.

IV. *Degenerate chain branching*, i.e., the decomposition of hydroperoxides to radicals, is the most important step in polymer oxidation. When the oxidation proceeds to some extent and hydroperoxide is accumulated, reaction (IV.1) will be the *main initiation reaction* in autoxidation processes. Depending on the circumstances, the thermal decomposition of hydroperoxides may proceed unimolecularly, bimolecularly, or with participation of an RH group. Thus, for examples, the decomposition of adjacent hydroperoxide groups formed according to reaction (6.1) proceeds bifunctionally and ten times faster than the unimolecular decomposition of isolated hydroperoxy groups. Induced decomposition, i.e., decomposition of hydroperoxides catalyzed by radicals and oxygen containing groups formed during the oxidation, may play an important role in the process. It is assumed that the presence of both carbonyl and carboxyl groups accelerates the decomposition; acid-catalyzed decomposition, however, favors the formation of molecular products ($f = 0$).

Hydroperoxides are oxidation intermediates of key importance. Their formation can be inhibited by "chain breaking" antioxidants which interfere with reaction (III.1). Their decomposition can also be influenced: some very important antioxidants act by decomposing the hydroperoxides to molecular products (peroxide decomposers). It is believed that sulfides, sulfoxides, thiosulfonates, selenides, tertiary amines, phosphines, phosphites, etc., are capable of decomposing hydroperoxides without the formation of free radicals, and thus function as stabilizers. The interaction of hydroperoxides with a thioether may be represented by:

$$\underset{R'}{\overset{R'}{\diagdown}}S + ROOH \rightarrow \underset{R'}{\overset{R'}{\diagdown}}S{=}O + ROH \qquad (6.2)$$

(Other methods of stabilization against oxidation will be treated in detail later.)

Reaction (IV.1) is very important not only because of its chain branching action but also with respect to the formation of oxidation products. The tertiary alkoxy radicals (RO·) produced in this reaction in polypropylene oxidation undergo β-scission (which leads to ketone formation) rather than participate in (IV.2) hydrogen abstraction:

$$\sim CH_2-\underset{\underset{O\cdot}{|}}{\overset{\overset{CH_3}{|}}{C}}-CH_2-\sim \;\rightarrow\; \sim CH_2-\underset{\underset{O}{\|}}{\overset{\overset{CH_3}{|}}{C}} \;+\cdot CH_2-\sim \qquad (6.3)$$

Besides ketone and water, many other products can be formed in step (IV.1), or in the reactions immediately following it. The formation of most of the volatile products (alcohols, acids, aldehydes) is connected to the hydroperoxide decomposition reaction.

V. Under most conditions, *unimolecular or bimolecular termination* of free radicals in polymer oxidation occurs almost exclusively by participation of peroxy radicals because the reaction of alkyl radicals with molecular oxygen, reaction (II), is very fast. However, at low oxygen pressures, termination reactions of the alkyl radicals are also important.

The molecularity of the termination reactions is highly disputed in polymer oxidation because there is no clear explanation of the unimolecular termination. Despite the lack of an unambiguous chemical description of this radical reaction, unimolecular termination must occur when solid polymers or polymer melts of high viscosity are oxidized. A bimolecular termination step is unlikely since it would depend upon two very small concentrations requiring polymer-polymer interaction in a medium of very high viscosity.

When polymer solutions or low viscosity polymer melts are oxidized, termination is assumed to involve reaction of two polymeric radicals. It is believed that this reaction involves the formation of a cyclic tetroxide intermediate which can rearrange and decompose to give a ketone, an alcohol, and oxygen:

$$2R-\underset{\underset{H}{|}}{\overset{\overset{R'}{|}}{C}}-OO\cdot \;\rightleftharpoons\; \underset{R}{\overset{R'}{\diagdown}}C\underset{H\cdots O}{\overset{O-O}{\diagup}}O \quad \underset{R}{\overset{R'}{\diagdown}}C \quad H \;\rightarrow\; \underset{R}{\overset{R'}{\diagdown}}C=O + O_2 + HO-\underset{\underset{R}{|}}{\overset{\overset{H}{|}}{C}}-R'$$

$$(6.4)$$

This mechanism provides an attractive explanation for the rapid termination of secondary peroxy radicals. It has been proposed that the molecular oxygen formed in reaction (6.4) is in an excited state (a so-called singlet oxygen) and emits light when it becomes deactivated.

6.2 EFFECT OF POLYMER PROPERTIES ON OXIDATION

The rate of diffusion of oxygen into the polymer and the solubility of gases are greater in *amorphous* regions than in *crystalline* domains. The crystalline cores are inaccessible to molecular oxygen. This implies that the rate of oxidation is greater in amorphous than in crystalline polymers, and in fact, this has been observed experimentally. The oxidation rate of polyethylene is inversely proportional to the degree of crystallinity. The mobility of radicals is larger in amorphous regions; the effect of crystallinity is absent when the oxidation is carried out in solution. Polymer melts, especially at low temperatures, often preserve parts of the crystalline order they had before melting.

It is observed that crystallinity increases with the degree of oxidation. This is attributed to the breaking of chains in amorphous regions; these chains then crystallize. Also, the diffusion coefficient of the material changes as a result of the morphological alterations; the presence of oxygen containing groups built into the polymer during oxidation affects diffusion rate.

It has long been known that the *tacticity* of polymers influences their ability to crystallize. Stereoregular polymers are often highly crystalline and should be more resistant to oxidation than the corresponding polymers having irregular structure. In polypropylene, however, the rate of oxidation of the stereoregular polymer is greater than that of the atactic material. Thus, atactic polypropylene oxidizes 20–40% slower than isotactic polypropylene. This seeming discrepancy has been attributed to a stereo-dependent oxidation reaction.

The *molecular weight* may affect the oxidizability of polymers. The effect of molecular weight is different for oxidation of bulk polymers and polymer solutions. The oxidation rate of bulk polystyrene and polypropylene is virtually independent of molecular weight. This may be explained by considering the rates of initiation and termination: increasing molecular weight decreases the rate of initiation, but the rate of termination also decreases. The net effect is that the oxidizability is virtually independent of molecular weight.

The oxidation of polymers in dilute solutions exhibits a different dependence on the polymer molecular weight from that found in the bulk oxidation. In general, the apparent reactivity and the rate of oxidation of polymers in solution decrease as the polymer molecular weight increases. However, in dilute solutions the mobility of radicals may, in some cases, not decrease with an increase in the viscosity of the medium.

The *chemical structure* of a polymer strongly affects its ability to resist oxidative degradation. It is known that the rate of degradation increases with increased chain branching. This is expected: "branchy" materials contain more tertiary hydrogen atoms than pure linear materials. As previously mentioned, tertiary hydrogen atoms are more vulnerable to radical attack than are primary or secondary hydrogen atoms. The slow termination of tertiary peroxy radicals may also contribute to the increased rate of oxidation.

The presence of double bonds in a polymer backbone affects the mechanism of oxidative degradation Free radicals can add to double bonds. Allylic hydrogen atoms readily react with free radicals to form resonance-stabilized radicals. Hydrogen abstraction would, of course, allow cross-linking reactions to occur.

Polymer films are often used in the study of photochemical oxidation; however, care must be taken when making measurements on polymer films because the *film thickness* greatly affects the degradation. Film thickness affects the rate and products of oxidation, the absorption of light, and the diffusion of oxygen. For example, the intensity of 253.7 nm radiation is reduced by 99% by a 0.01 mm thick polystyrene film. The overall effect of film thickness is that the applied stress (e.g., irradiation) and hence the rate of degradation are not constant throughout the thickness of the film.

In general, the rate of oxidation decreases with increasing film thickness. Thick films display longer induction periods and slower rates of oxidation than thinner films. The oxidation of thick films depends on oxygen pressure, and oxidation occurs mostly on the film surface; the oxidation reaction is, at this point, diffusion controlled. Below a critical film thickness, the rate of oxidation is not diffusion controlled; rather, it depends on the chemistry of the oxidation reactions. The critical film thickness is not a constant; on the contrary, it depends on the experimental conditions at which measurements are made.

In Figure 6.3 anisothermal TG and DTA curves of isotactic polypropylene films (thickness 0.05 and 1 mm) are shown. For the thin sample, the small endotherm peak of semicrystalline melting point (T_1) is followed by one sharp exotherm peak (T_2') of oxidation. For the thick sample, however, the exotherm peak (T_2) is followed by further exotherm peaks (T_3, T_4, T_5); this results from the slowing down and delay of oxidation reactions because of diffusion control. Also, TG curves show that only the oxidation of the thin sample is independent of the rate of the transport processes, i.e., the diffusion rate of oxygen into the film and of volatile products outward.

Various compounds such as *additives* and *impurities* are present in commercially prepared polymers. These compounds, which include catalyst residues and processing aids, may affect the oxidation of polymers. Metallic impurities (generally catalyst residues) diminish the induction period and accelerate the oxidation. Metals in low valence states may retard the initial oxidation reactions, however, they later accelerate oxidation by decomposing hydroperoxides formed in the early stages of oxidation.

Oxidation products may affect the rate of oxidation of the parent polymer. Since many of the first formed oxidation products are more easily oxidized than the parent polymer, autocatalysis is observed. Hydroperoxides, commonly the major product of oxidative degradation, are potentially powerful initiators of further degradation. Hydroperoxides may be decomposed thermally, photo-

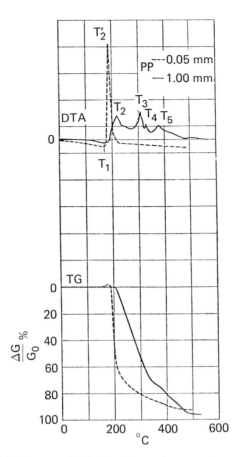

Figure 6.3. TG and DTA curves of isotactic polypropylene samples measured under oxygen atmosphere. Film thickness: 1 mm and 0.05 mm (dotted lines).

chemically, or catalytically to form free radicals which may attack the parent polymer and cause further degradation. Carbonyl groups are assumed to catalyze photooxidation of polymers. In some cases, oxidation products retard further polymer degradation.

6.3 KINETICS OF POLYOLEFIN THERMAL OXIDATION

The general scheme of polymer oxidation (Figure 6.1) can be further simplified when our intention is merely the kinetic description of the main characteristics of the process. Such characteristics are the time dependence of oxygen absorption, and the temperature, pressure, and concentration dependence of the oxidation rate. Of course, there are many other important features of polymer oxidation such as the change of MWD, the formation of volatile products, and the change of polymer composition due to the built-in oxygen containing groups. A scheme which would allow the kinetic treatment of all processes determining these features would include many more steps than that shown in Figure 6.1, and the reactions would be chemically much more specific. Such a treatment is, however, beyond the framework of this text.

An allowable simplification is the assumption that only the $RO_2 \cdot$ radicals and (at low and medium oxygen pressures) the $R \cdot$ radicals are involved in rate-determining steps and that the reactions of the low molecular weight, mobile radicals ($R' \cdot$ initiator fragment, $HO \cdot$, and $HO_2 \cdot$) are very fast. We may also assume that the alkoxy ($RO \cdot$) radicals are rapidly converted to alkyl radicals. Thus we built reactions (I.3), (I.4), (IV.2), and (IV.3) of Figure 6.1 into the other reactions and used them only as stoichiometric factors, listed in the parentheses in Figures 6.4 and 6.5. The scheme in Figure 6.5 was constructed with a further simplification: at high oxygen pressures the rate of the radical conversion reaction (II) is high; i.e., the $R \cdot$ alkyl radicals are very rapidly converted to $RO_2 \cdot$ peroxy radicals and do not participate in the rate-determining steps.

I. *Initiation*:

$$RH + O_2 \; (+ RH) \longrightarrow 2R^{\cdot} + H_2O_2 \tag{I.1}$$
$$\text{Initiator} \; (+ 2f'RH) \longrightarrow 2f'R^{\cdot} + 2f'R'H \tag{I.2}$$

II. *Radical conversion (stabilization)*:

$$R^{\cdot} + O_2 \longrightarrow RO_2^{\cdot} \tag{II}$$

III. *Chain propagation*:

$$RO_2^{\cdot} + RH \longrightarrow R^{\cdot} + RO_2H \tag{III.1}$$
$$RO_2^{\cdot} + RH \longrightarrow R^{\cdot} + \text{products} \tag{III.2}$$

IV. *Degenerate chain branching*:

$$RO_2H \; (+ 2fRH) \longrightarrow 2fR^{\cdot} + fROH + (1-f)RO + H_2O \tag{IV}$$

V. *Termination*:

$$R^{\cdot} \longrightarrow \text{products} \tag{V.1}$$
$$RO_2^{\cdot} \longrightarrow \text{products} \tag{V.2}$$
$$R^{\cdot} + R^{\cdot} \longrightarrow \text{products} \tag{V.3}$$
$$R^{\cdot} + RO_2^{\cdot} \longrightarrow \text{products} \tag{V.4}$$
$$RO_2^{\cdot} + RO_2^{\cdot} \longrightarrow \text{products} \tag{V.5}$$

Figure 6.4. Simplified scheme of polymer oxidation at low and medium oxygen partial pressures.

I. *Initiation*:

$$RH + O_2 (+ RH + 2O_2) \longrightarrow 2RO_2^{\cdot} + H_2O_2 \tag{I.1}$$

$$\text{Initiator} (+ 2f'RH + 2f'O_2) \longrightarrow 2f'RO_2^{\cdot} + 2f'R'H \tag{I.2}$$

III. *Chain propagation*:

$$RO_2^{\cdot} + RH (+ O_2) \longrightarrow RO_2^{\cdot} + RO_2H \tag{III.1}$$

$$RO_2^{\cdot} + RH (+ O_2) \longrightarrow RO_2^{\cdot} + \text{products} \tag{III.2}$$

IV. *Degenerate chain branching*:

$$RO_2H (+ 2fRH + 2fO_2) \longrightarrow 2fRO_2^{\cdot} + fROH + (1-f)RO + H_2O \tag{IV}$$

V. *Termination*:

$$RO_2^{\cdot} \longrightarrow \text{products} \tag{V.2}$$

$$RO_2^{\cdot} + RO_2^{\cdot} \longrightarrow \text{products} \tag{V.5}$$

Figure 6.5. Simplified scheme of polymer oxidation at high oxygen pressures.

Again, the amount of O_2 consumed in reaction (II) is included as a stoichiometric factor, listed in the parentheses. Thus, the difference between the schemes is that Figure 6.4 involves $R\cdot$ and $RO_2\cdot$ radicals (low and medium O_2 pressures) in the rate-determining steps, while in Figure. 6.5 only $RO_2\cdot$ radicals (high O_2 pressure) are involved.

Further simplifications are included in the following kinetic treatment. (1) We assume the concentration of RH, i.e., the C—H bonds participating in the oxidation, to be constant. This is a very good approximation at the beginning of the reaction, but applicable only up to the period of maximum rate. (2) We assume that a steady state is reached very soon after the onset of the oxidation. This is a usual and repeatedly justified assumption of the kinetics of chain reactions. (3) We take only either unimolecular (solid polymer) or bimolecular (solution or melt) radical termination reactions into account; furthermore, when working with the scheme in Figure 6.4, we assume that the termination rate constants are correlated: $k_{51} = k_{52} = k_5$ in the unimolecular case, and $k_{53} = k_{54}/2 = k_{55} = k_5$ in the bimolecular case. These simplifications are employed to obtain tractable expressions, and the implicit error is usually insignificant.

(A) Let us first consider the *time dependence of autoxidation* for a solid polymer sample at low or medium O_2 pressure, i.e., using the scheme in Figure 6.4 with unimolecular termination. The applied notations are the following:

$$[R\cdot] = R \quad [RO_2\cdot] = X \quad [RO_2H] = Y \quad [O_2]_{abs} = Z$$

$$[RH] = c = \text{constant} \quad [O_2] = kp = \text{constant} \quad [\text{initiator}] = I \tag{6.5}$$

where Z stands for the amount of oxygen absorbed by the polymer and p is the O_2 pressure.

The rates for the individual steps are as follows:

$$W_{11} = k_{11}''' ckp = k_{11}'' c = k_{11}' p = k_{11} \quad W_{12} = k_{12} I$$

$$W_2 = k_2'' Rkp = k_2' Rp = k_2 R$$

$$W_{31} = k_{31}' cX = k_{31} X \quad W_{32} = k_{32}' cX = k_{32} X \tag{6.6}$$

$$W_4 = k_4 Y$$

$$W_{51} = k_5 R \quad W_{52} = k_5 X \quad W_{53} = k_5 R^2 \quad W_{54} = 2k_5 RX \quad W_{55} = k_5 X^2$$

With these individual rates, the process can be described by the following differential equations:

$$\frac{dR}{dt} = 2W_{11} - W_2 + W_{31} + W_{32} + 2fW_4 - W_{51} \tag{6.7}$$

$$\frac{dX}{dt} = W_2 - W_{31} - W_{32} - W_{52} \tag{6.8}$$

$$\frac{dY}{dt} = W_{31} - W_4 \tag{6.9}$$

$$\frac{dZ}{dt} = W_{11} + W_2 \tag{6.10}$$

Under steady-state conditions, dR/dt and dX/dt are nearly equal to zero. From Eq. (6.8) we obtain:

$$W_{52} = W_2 - (W_{31} + W_{32}) \tag{6.11}$$

or, using Eq. (6.6):

$$X = \frac{k_2}{k_{31} + k_{32} + k_5} R \tag{6.12}$$

Substituting Eq. (6.11) into Eq. (6.7), we obtain:

$$W_{51} + W_{52} = 2(W_{11} + fW_4) \tag{6.13}$$

or, using Eq. (6.6):

$$X + R = \frac{2}{k_5}(k_{11} + fk_4 Y) \tag{6.14}$$

From Eqs. (6.12) and (6.14) we obtain for the individual radical concentrations:

$$X = \alpha \frac{2}{k_5}(k_{11} + fk_4 Y) \tag{6.15}$$

and

$$R = (1 - \alpha)\frac{2}{k_5}(k_{11} + fk_4 Y) \tag{6.16}$$

where

$$\alpha = \frac{k_2}{k_2 + k_{31} + k_{32} + k_5} \tag{6.17}$$

Now we can solve differential equation (6.9) for the time dependence of Y. Using Eqs. (6.6), (6.15), and (6.16), we obtain from Eq. (6.9):

$$\frac{dY}{dt} = a - bY \tag{6.18}$$

where

$$a = \frac{2k_{11}k_{31}}{k_5}\alpha \tag{6.19}$$

and

$$b = \left(1 - \frac{2fk_{31}}{k_5}\alpha\right)k_4 \tag{6.20}$$

Integration of the nonhomogeneous linear differential equation (6.18) gives the following result:

$$Y = e^{-bt}\left(C + \frac{a}{b}e^{bt}\right) \tag{6.21}$$

where the value of the C integration constant can be obtained from the initial conditions:

$$t = 0 \quad Y = 0 \quad C = -\frac{a}{b} \tag{6.22}$$

Hence, our result for the time dependence of the hydroperoxide concentration is:

$$Y = \frac{a}{b}(1 - e^{-bt}) \tag{6.23}$$

which is represented in Figure 6.6.

Using Eq. (6.23) in the expression (6.16) of R, we can calculate the rate of oxygen absorption (W_Z) from Eq. (6.10):

$$W_Z = \frac{dZ}{dt} = k_{11}\left[(1 + \nu) - (\nu - \nu_0)e^{-bt}\right] \tag{6.24}$$

where ν_0 and ν are the kinetic chain lengths of the oxidation. At the beginning of the process:

$$W_{Z,0} = k_{11}(1 + \nu_0) \quad \text{with} \quad \nu_0 = (1-\alpha)\frac{2k_2}{k_5} \tag{6.25}$$

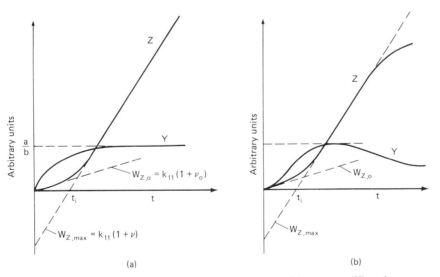

Figure 6.6. (a) Calculated and (b) experimental hydroperoxide content (Y) and oxygen absorption (Z) curves in polyolefin oxidation. (Z and Y have different units.)

At the maximum rate of oxidation:

$$W_{Z,\max} = k_{11}(1 + \nu) \quad \text{with} \quad \nu = \nu_0 \left(1 + \frac{fk_4}{k_{11}} \cdot \frac{a}{b}\right) = \nu_0 (1 + \mu) \qquad (6.26)$$

The second term in the parenthesis (μ) shows the measure of autoacceleration of the process. In the period of maximum oxidation rate, the hydroperoxide concentration [see Eq. (6.23)] approaches its maximum value:

$$Y_{\max} = \frac{a}{b} \qquad (6.27)$$

Thus,

$$\mu = \frac{fk_4 Y_{\max}}{k_{11}} \qquad (6.28)$$

is nothing more than the ratio of the maximum rate of initiation by degenerate chain branching to the rate of primary initiation.

Integration of Eq. (6.24) supplies the time dependence of the amount of oxygen absorbed by the polymer:

$$Z = k_{11} \left[(1 + \nu)t - \frac{\nu - \nu_0}{b}(1 - e^{-bt})\right] \qquad (6.29)$$

which is also represented in Figure 6.6. The length of the induction period, t_i, can be calculated from Eq. (6.29):

$$t_i = \frac{1}{b} \cdot \frac{\nu - \nu_0}{1 + \nu} \qquad (6.30)$$

In Figure 6.6, the time dependence of the hydroperoxide concentration and of the absorbed oxygen in a real oxidation system is represented as well. The comparison of these curves with the calculated ones shows two main differences. (1) The real hydroperoxide accumulation begins only after a short time lag; the calculated hydroperoxide accumulation, however, begins without delay. The reason for this is that we assumed steady-state conditions from $t = 0$ which is, of course, only an approximation. (2) The real hydroperoxide curve shows a decrease after reaching the maximum. Similarly, on the real oxygen absorption curve the rate decreases again after reaching its maximum value. The reason for this deviation is that in our derivation we assumed the RH concentration to be constant. However, this is a good approximation only at the beginning of the oxidation. On the other hand, some properties of the polymer change with the progress of oxidation, e.g., its oxygen permeability, crystallinity, etc., which cause a change of the kinetic parameters too.

(B) The *pressure dependence of autoxidation* is an important problem in polymer oxidation. We will discuss it in the following material for the case of a polyethylene melt; thus we must again use the scheme in Figure 6.4, but with bimolecular termination. We apply the same notation as before, and differential equations (6.9) and (6.10) are unchanged. The difference between cases A and B is only in the rates of the termination reactions of the radicals. The differential equations for the radicals are now:

$$\frac{dR}{dt} = 2W_{11} - W_2 + W_{31} + W_{32} + 2fW_4 - 2W_{53} - W_{54} \tag{6.31}$$

$$\frac{dX}{dt} = W_2 - W_{31} - W_{32} - W_{54} - 2W_{55} \tag{6.32}$$

Under steady-state conditions ($dR/dt = 0$, $dX/dt = 0$), the sum of Eqs. (6.31) and (6.32) is:

$$W_{53} + W_{54} + W_{55} = W_{11} + fW_4 \tag{6.33}$$

or, using Eq. (6.6):

$$X + R = \left(\frac{k'_{11}p + fk_4 Y}{k_5}\right)^{1/2} \tag{6.34}$$

Substituting Eq. (6.34) into Eq. (6.32), and using the rate expressions from Eq. (6.6), we obtain:

$$\frac{X}{R} = \frac{k'_2 p}{k_{31} + k_{32} + 2k_5 (X + R)} \tag{6.35}$$

From Eqs. (6.34) and (6.35) we obtain for the individual radical concentrations:

$$X = \beta \left(\frac{k'_{11}p + fk_4 Y}{k_5}\right)^{1/2} \tag{6.36}$$

and

$$R = (1 - \beta) \left(\frac{k'_{11}p + fk_4 Y}{k_5}\right)^{1/2} \tag{6.37}$$

where

$$\beta = \frac{k_2'p}{k_2'p + k_{31} + k_{32} + 2k_5(X+R)} \tag{6.38}$$

Compared to the propagation rate constants, the $2k_5(X+R)$ term is negligible; thus,

$$\beta = \frac{k_2'p}{k_2'p + k_{31} + k_{32}} \tag{6.39}$$

Let us first consider the pressure dependence of the *initial rate* of oxygen absorption. At the beginning of the oxidation there is no hydroperoxide in the polymer, i.e., $Y_0 = 0$; thus the value of $W_{Z,0}$ can be obtained from Eq. (6.10) by using

$$R_0 = (1 - \beta) \left(\frac{k_{11}'p}{k_5} \right)^{1/2} \tag{6.40}$$

and the rate expressions from Eq. (6.6):

$$W_{Z,0} = k_{11}'p + k_2'p\,(1 - \beta) \left(\frac{k_{11}'p}{k_5} \right)^{1/2} \tag{6.41}$$

Neglecting the first term and substituting β from Eq. (6.39), we obtain:

$$W_{Z,0} = (k_{31} + k_{32}) \left(\frac{k_{11}'}{k_5} \right)^{1/2} \frac{p^{3/2}}{\dfrac{k_{31} + k_{32}}{k_2'} + p} \tag{6.42}$$

or in a linear form:

$$\frac{p^{3/2}}{W_{Z,0}} = \frac{\dfrac{k_{31} + k_{32}}{k_2'} + p}{(k_{31} + k_{32}) \left(\dfrac{k_{11}'}{k_5} \right)^{1/2}} \tag{6.43}$$

Figure 6.7 shows the experimental data for the initial rate of oxygen absorption measured on molten high pressure polyethylene samples at various temperatures. Application of Eq. (6.43) provides a linear relationship which gives satisfactory straight lines, despite the fact that the initial rate can be determined or reproduced only with limited accuracy.

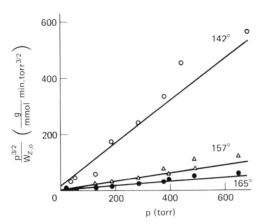

Figure 6.7. Linearization of the pressure dependence of the initial rate of oxygen absorption, according to Eq. (6.43). Molten samples of high pressure polyethylene. (Reprinted, by permission of Pergamon Press, from reference 6.15.)

During the period of *maximum oxidation rate*, the primary initiation can be neglected compared to initiation by degenerate chain branching; thus, Eq. (6.36) can be written in the following form:

$$X_{max} \approx \beta \left(\frac{f k_4 Y_{max}}{k_5} \right)^{1/2} \tag{6.44}$$

On the other hand, during this period $dY/dt = 0$, i.e., using Eq. (6.6), we have from Eq. (6.9):

$$X_{max} = \frac{k_4 Y_{max}}{k_{31}} \tag{6.45}$$

From Eqs. (6.44) and (6.45) we obtain:

$$Y_{max}^{1/2} = \frac{k_{31}}{k_4} \beta \left(\frac{f k_4}{k_5} \right)^{1/2} \tag{6.46}$$

Hence, using Eq. (6.37), the value of R in this period is:

$$R_{max} = (1 - \beta) \beta \frac{k_{31}}{k_5} f \tag{6.47}$$

The maximum rate of oxygen absorption can be obtained from Eq. (6.10), using Eq. (6.47) and the rate expressions from Eq. (6.6):

$$W_{Z,max} = k'_{11}p + k'_2 p(1 - \beta)\beta \frac{k_{31}}{k_5}f \tag{6.48}$$

Neglecting the first term and substituting β from Eq. (6.39), we obtain:

$$W_{Z,max} = (k_{31} + k_{32})\frac{k_{31}}{k_5}f \frac{p^2}{\left(\dfrac{k_{31} + k_{32}}{k'_2} + p\right)^2} \tag{6.49}$$

or in a linear form:

$$\frac{p}{(W_{Z,max})^{1/2}} = \frac{\dfrac{k_{31} + k_{32}}{k'_2} + p}{\left[(k_{31} + k_{32})\dfrac{k_{31}}{k_5}f\right]^{1/2}} \tag{6.50}$$

The linear pressure dependence of experimental data on the maximum rate of oxygen absorption of molten high pressure polytheylene samples is shown in Figure 6.8. It can be seen that the pressure dependence of the maximum rate is well described by Eq. (6.49) at the three temperatures examined.

(C) The *polymer concentration dependence of autoxidation* is important when studying oxidation in solution. We will investigate it in relation to polyethylene oxidation in trichlorobenzene (TCB) solution at high oxygen pressure (1 atm), i.e., using the scheme in Figure 6.5 with bimolecular termination. We apply the same notation as before. The differential equations of the system are as follow:

$$\frac{dX}{dt} = 2W_{11} + 2fW_4 - 2W_{55} \tag{6.51}$$

$$\frac{dY}{dt} = W_{31} - W_4 \tag{6.52}$$

$$\frac{dZ}{dt} = 3W_{11} + W_{31} + W_{32} + 2fW_4 \tag{6.53}$$

Under steady-state conditions dX/dt is nearly equal to zero, thus from Eq. (6.51) we obtain:

$$X = \left(\frac{k''_{11}c + fk_4 Y}{k_5}\right)^{1/2} \tag{6.54}$$

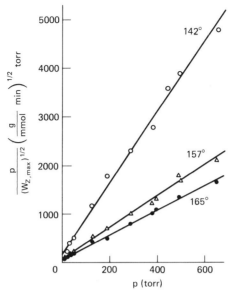

Figure 6.8. Linearization of the pressure dependence of the maximum rate of oxygen absorption, according to Eq. (6.50). Molten samples of high pressure polyethylene. (Reprinted, by permission of Pergamon Press, from reference 6.15.)

Let us first consider the polymer concentration dependence of the *initial rate* of oxidation. At $t = 0$ there is no hydroperoxide in the system, i.e., $Y_0 = 0$; thus,

$$X_0 = \left(\frac{k_{11}'' c}{k_5}\right)^{1/2} \tag{6.55}$$

and the value of $W_{Z,0}$, using Eq. (6.53) with the rate expressions from Eq. (6.6), is

$$W_{Z,0} = 3k_{11}'' c + (k_{31}' + k_{32}') c \left(\frac{k_{11}'' c}{k_5}\right)^{1/2} \tag{6.56}$$

or, because the second term on the right-hand side of Eq. (6.56) is much higher than the first one, we may write:

$$W_{Z,0} = (k_{31}' + k_{32}') \left(\frac{k_{11}''}{k_5}\right)^{1/2} c^{3/2} \tag{6.57}$$

i.e., the initial oxidation rate is directly proportional to the 3/2 power of the polymer concentration.

During the period of *maximum oxidation rate*, the primary initiation can be neglected; thus Eq. (6.54) can be written as follows:

$$X_{max} = \left(\frac{fk_4 Y_{max}}{k_5}\right)^{1/2} \tag{6.58}$$

and, because in this period $dY/dt = 0$, we have from Eq. (6.52):

$$X_{max} = \frac{k_4}{k'_{31}c} Y_{max} \tag{6.59}$$

From Eqs. (6.58) and (6.59) we obtain:

$$Y_{max} = \frac{fk'^2_{31}}{k_4 k_5} c^2 \tag{6.60}$$

The maximum rate of oxidation is then, according to Eq. (6.53),

$$W_{Z,\,max} = 3k''_{11}c + \left[(k'_{31} + k'_{32})\frac{k_4}{k'_{31}} + 2fk_4\right]\frac{fk'_{31}}{k_4 k_5}c^2 \tag{6.61}$$

or, if we neglect the first term,

$$W_{Z,\,max} = [k'_{31}(1 + 2f) + k'_{32}]\frac{fk'_{31}}{k_5}c^2 \tag{6.62}$$

i.e., the maximum rate of oxidation is directly proportional to the square of the polymer concentration. Experimental data on $W_{Z,\,max}$ obtained for autoxidation of high pressure polyethylene in trichlorobenzene solution at 160°C are presented vs $[RH]^2$ in Figure 6.9. As can be seen, Eq. (6.62) fits the data well. Extrapolation to the polymer concentration of the melt (33 mol/liter), by using the slope of the line obtained from the solution, leads to the value $W_{Z,\,max} = 0.135$ mol/liter/min which is in good agreement with the experimental value (0.107). This indicates that the mechanisms of autoxidation of PE in TCB solution and in the molten phase are not very different at 160°C; thus we may rightly assume bimolecular termination in the melt also.

(D) The *initiator concentration dependence of oxidation* can be simply investigated only in the case of a dissolved polymer. In the following, we will discuss the initiated oxidation of high pressure polyethylene dissolved in trichlorobenzene, with dicumyl peroxide as initiator. We apply the scheme in

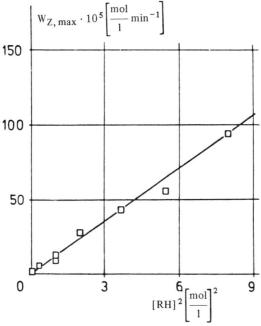

Figure 6.9. Plot according to Eq. (6.62) of maximum rate of autoxidation of high pressure polyethylene in trichlorobenzene solution at 160°C, 760 torr O_2 pressure. (Reprinted, by permission of Applied Science Publishers Ltd., from reference 6.13.)

Figure 6.5 with bimolecular termination and neglect the (I.1) thermal initiation as insignificant compared to the initiation by dicumyl peroxide.

$$\frac{dX}{dt} = 2f'W_{12} + 2fW_4 - 2W_{55} \tag{6.63}$$

$$\frac{dY}{dt} = W_{31} - W_4 \tag{6.64}$$

$$\frac{dZ}{dt} = 2f'W_{12} + W_{31} + W_{32} + 2fW_4 \tag{6.65}$$

Under steady-state conditions we obtain from Eq. (6.63):

$$X = \left(\frac{f'k_{12}I + fk_4Y}{k_5}\right)^{1/2} \tag{6.66}$$

In the presence of sufficient amount of initiator, the rate of acceleration caused by the degenerate chain branching is not as important as in the case of autoxidation. It is, therefore, enough to study the initiator concentration dependence of the initial oxidation rate. At the beginning of the reaction, $Y_0 = 0$; thus

$$X_0 = \left(\frac{f'k_{12}I_0}{k_s}\right)^{1/2} \tag{6.67}$$

and the value of $W_{Z, 0}$, from Eq. (6.65),

$$W_{Z, 0} = 2f'k_{12}I_0 + (k'_{31} + k'_{32})\left(\frac{f'k_{12}I_0}{k_s}\right)^{1/2} c \tag{6.68}$$

or in a linear form,

$$\frac{W_{Z, 0}}{I_0} = 2f'k_{12} + (k'_{31} + k'_{32})\left(\frac{f'k_{12}}{k_s}\right)^{1/2} \frac{c}{I_0^{1/2}} \tag{6.69}$$

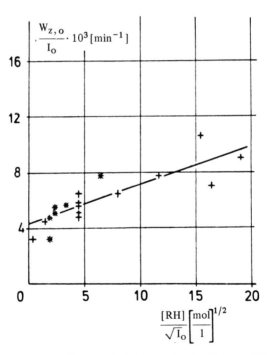

Figure 6.10. Plot according to Eq. (6.69) of initial rate of oxidation of high pressure polyethylene in trichlorobenzene solution, initiated by dicumyl peroxide, at 120°C, 760 torr O_2 pressure. (Reprinted, by permission of Applied Science Publishers Ltd., from reference 6.13.)

The plot of experimental data according to Eq. (6.69) is shown in Figure 6.10. Although the plot consists of data obtained by two different experimental techniques (+: constant I_0 with varying c; *: varying I_0 with constant c), the agreement with Eq. (6.69) is satisfactory.

6.4 SOME DIFFERENCES IN THE OXIDATION OF POLYETHYLENE AND POLYPROPYLENE

A number of similar features can be observed in the thermal oxidation of different polyolefins. The overall oxygen absorption and the formation of end products are autocatalytic; the concentration of polymer hydroperoxide (the intermediate product of oxidation) passes through a maximum. These similarities are well illustrated by the Z and Y curves for high pressure (low density) polyethylene (PE) and isotactic polypropylene (PP) in Figure 6.11. As can be seen, the character of these curves is similar, only the quantitative aspects (rate, etc.) are different. There are, however, some marked differences between the two polymers with respect to the structure of hydroperoxides formed, fragmentation of the polymer, formation and composition of volatile materials, etc.

Comparison of the infrared spectra of oxidized PE and PP shows characteristic differences. Figure 6.12 displays spectra of PE and PP films with approximately identical oxygen content in the ranges of 2900-3700 and 1600-1800 cm^{-1}, respectively.

In the range of 2900-3700 cm^{-1} where hydroxyl groups with different chemical structures absorb, PP shows a significant band with a maximum at 3380 cm^{-1}, while the absorption of PE in this range is much lower and lacks a

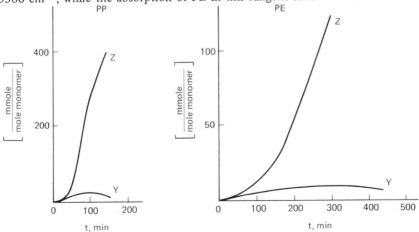

Figure 6.11. Amount of absorbed oxygen (Z) and of hydroperoxides (Y) as a function of oxidation time at 140°C, 760 torr O$_2$ pressure (PE and PP film samples). (Reprinted, by permission of Pergamon Press, from reference 6.12.)

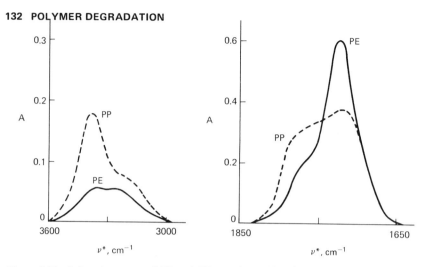

Figure 6.12. Infrared spectra of PE and PP samples oxidized to approximately the same oxygen content. (Reprinted, by permission of Pergamon Press, from reference 6.12.)

sharp peak. This difference can be attributed mainly to the formation of hydroperoxide sequences in the PP sample [see reaction (6.1)] for which the absorption coefficient is much higher than for the isolated hydroperoxide groups formed in PE.

Figure 6.13 shows the hydroperoxide content as a function of the amount of oxygen absorbed (Z); i.e., it compares Y values belonging to the same extent of oxidation. It is apparent that, in the initial stage of the reaction, the number of hydroperoxide groups per molecule of absorbed oxygen is independent of both the type of the polymer and the reaction conditions and has a value close to unity. This points to almost quantitative formation of hydroperoxides from the absorbed oxygen.

At higher conversion, hydroperoxide content depends on the reaction conditions and on the nature of the polymer. With both polymers, over the range of temperature and pressure studied, hydroperoxide concentration was strongly dependent on the pressure but scarcely, or not at all, dependent on the temperature. Under comparable conditions (i.e., at nearly identical overall rates of oxidation, under 760 torr oxygen pressure, at 160°C for PE and 130°C for PP), the amount of hydroperoxide assigned to an identical amount of total oxygen absorption is always higher in PP. Thus, the other cause of the difference in the OH bond shown in Figure 6.12 is quantitative.

Extinction curves are also quite different in the range of 1600–1800 cm⁻¹ (see Figure 6.12). In PE mainly keto and carboxyl groups can be found (1700–1740 cm⁻¹), while in PP the formation of considerable amounts of other carbonyl compounds is observed. The total amount of carbonyl formed from 1 mole of absorbed oxygen is also different in the two polymers (Figure 6.14). In

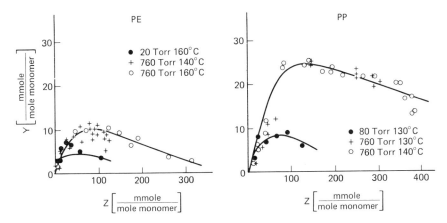

Figure 6.13. Hydroperoxide content (Y) fo PE and PP samples as a function of the amount of absorbed oxygen (Z) under various oxidation conditions. (Reprinted, by permission of Pergamon Press, from reference 6.12.)

the case of PE a larger fraction of the absorbed oxygen (approximately 1.5 times as much as in PP) is incorporated as carbonyl.

The ratio of incorporation appears to be independent of the reaction conditions. In the thermal oxidation of PE, the oxygen absorbed is approximately half in the solid and half in the volatile phase. In the case of PP, a smaller proportion of the absorbed oxygen is incorporated in the form of carbonyl, while

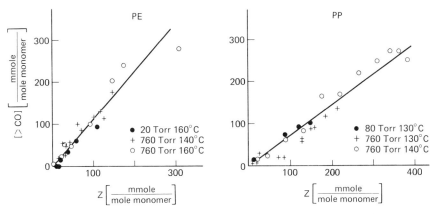

Figure 6.14. Carbonyl content ($[>CO]$) of PE and PP samples as a function of the amount of absorbed oxygen (Z) under various oxidation conditions. (Reprinted, by permission of Pergamon Press, from reference 6.12.)

most is built in partly in the form of higher hydroperoxide content and partly in the volatile products which are formed in higher amounts than in the case of PE.

Figure 6.15 depicts the amount of the volatile C_2-C_4 organic oxocompounds evolved during degradation as a function of conversion. It can be seen that the formation of volatile products from identical amounts of absorbed oxygen is lower for PE than PP. This phenomenon may be explained by the different structures of polymeric hydroperoxides formed as intermediate products from PP and PE during oxidation. In PP oxidation, the formation of hydroperoxide sequences, mainly of dihydroperoxides, can be observed. During bifunctional decomposition of these groups, complex transition states with several active centers are formed which lead to end products containing low molecular weight oxocompounds.

In the thermal oxidation of PE and PP, the most remarkable difference can be observed in the changes of molecular weights of the two polymers. Figure 6.16 shows the average number of scissions related to one original polymer molecule as a function of conversion.

The s value, determined from molecular weight measurements, is characteristic of scissions involving considerable molecular weight decrease. (The change of molecular weight distribution during oxidation of polypropylene was shown in

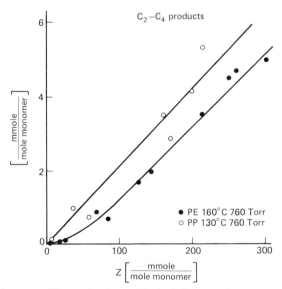

Figure 6.15. Amount of low molecular weight (volatile) organic oxo compounds (acetaldehyde, acetone, propionaldehyde, butyraldehyde, etc.) as a function of the amount of oxygen (Z) absorbed by the polymer. (Reprinted, by permission of Pergamon Press, from reference 6.12.)

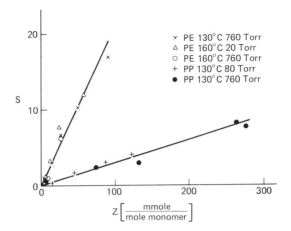

Figure 6.16. Average number of chain scissions leading to molecular weight decrease ($s = P_{n,0}/P_n - 1$) as a function of the extent of oxidation (Z). (Reprinted, by permission of Pergamon Press, from reference 6.12.)

Figure 2.24.) Since the formation of new chain ends usually leads to the formation of carboxyl groups (according to experience, one from each scission), the acidity of the oxidized polymer changes parallel with s and shows the same dependence on Z.

The number of scissions is much higher (sixfold) for identical degrees of oxidation in PE than in PP. The number of scissions in both polymers depends only on the amount of oxygen absorbed and is independent of the temperature and oxygen pressure.

This marked difference in the average number of scissions of identically oxidized PE and PP may be due to various reasons. Two possible causes are the formation of different hydroperoxide structures in the two polymers and different possibilities for bimolecular termination of macroradicals because of deviation in the physical characteristics of the two polymers.

REFERENCES

6.1. Bawn, C. E. H. and Chaudri, S. A. Autoxidation of atactic polypropylene in solution. *Polymer* 9:113, 123 (1968).

6.2. Beachell, H. C. and Beck, D. L. Thermal oxidation of deuterated polypropylenes. *J. Polymer Sci.* A3:457 (1965).

6.3. Bolland, J. L. Kinetic studies in the chemistry of rubber and related materials. VI. The benzoyl peroxide catalyzed oxidation of ethyl linoleate. *Trans. Faraday Soc.* 44:669 (1948).

6.4. Chien, J. C. W. Polymer reactions. III. Structure of polypropylene hydroperoxide. *J. Polymer Sci.* A1(6):381 (1968).

6.5. Chien, J. C. W. and Boss, C. R. Polymer reactions. V. Kinetics of autoxidation of polypropylene. *J. Polymer Sci.* **A1**(5):3091 (1967).

6.6. Denisov, Ye. T. Theoretical aspects of the assessment of oxidative processes in polymers under natural conditions. *Polymer Science USSR* **21**:577 (1979).

6.7. Emanuel, N. M., Denisov, Ye. T. and Maizus, Z. K. *Liquid-Phase Oxidation of Hydrocarbons.* New York: Plenum Press, 1967.

6.8. Hawkins, W. L. Oxidative degradation of high polymers. In (Tipper, C. F. H., ed.) *Oxidation and Combustion Reviews*, Vol. I. New York: Elsevier, 1965.

6.9. Holmström, A. and Sörvik, E. M. Thermal degradation of polyethylene in a nitrogen atmosphere of low oxygen content. *J. Appl. Polymer Sci.* **18**:761, 779 (1974).

6.10. Iring, M., Kelen, T. and Tüdős, F. Study of thermal oxidation of polyolefins. II. Effect of layer thickness on the rate of oxidation in the melt phase. *European Polymer J.* **11**:631 (1975).

6.11. Iring, M., László-Hedvig, S., Kelen, T., Tüdős, F., Füzes, L., Samay, G. and Bodor, G. Study of thermal oxidation of polyolefins. VI. Change of molecular weight distribution in the thermal oxidation of polyethylene and polypropylene. *J. Polymer Sci.: Symposium* **57**:55 (1976).

6.12. Iring, M., László-Hedvig, S., Barabás, K., Kelen, T. and Tüdős, F. Study of thermal oxidation of polyolefins. IX. Some differences in the oxidation of polyethylene and polypropylene. *European Polymer J.* **14**:439 (1978).

6.13. Iring, M., Kelen, T. and Tüdős, F. Initiated oxidation and auto-oxidation of polyethylene in trichlorobenzene solution. *Polymer Degradation and Stability* **1**:297 (1979).

6.14. Iring, M., Tüdős, F., Fodor, Zs. and Kelen, T. Study of thermal oxidation of polyolefins. X. Correlation between the formation of carboxyl groups and scission in the oxidation of polyethylene in the melt phase. *Polymer Degradation and Stability* **2**:143 (1980).

6.15. Kelen, T., Iring, M. and Tüdős, F. Study of thermal oxidation of polyolefins. III. Pressure and temperature dependence of the rate of oxygen absorption. *European Polymer J.* **12**:35 (1976).

6.16. Lundberg, W. O. *Autoxidation and Anti-Oxidants.* New York: Interscience, 1962.

6.17. Mayo, F. R. Free-radical autoxidations of hydrocarbons. *Accounts Chem. Research* **1**:193 (1968).

6.18. Niki, E., Decker, C. and Mayo, F. R. Aging and degradation of polyolefins. I. Peroxide-initiated oxidations of atactic polypropylene. *J. Polymer Sci.–Chem.* **11**:2813 (1973).

6.19. Rapoport, N. Ya. and Shlyapnikov, Yu. A. The interconnection between the structural-mechanical and chemical phenomena accompanying oxidative degradation processes in oriented polypropylene. *Polymer Science USSR* **17**:847 (1975).

6.20. Reich, L. and Stivala, S. S. *Autoxidation of Hydrocarbons and Polyolefins. Kinetics and Mechanisms.* New York: Marcel Dekker, 1969.

6.21. Scott, G. *Atmospheric Oxidation and Antioxidants.* Amsterdam: Elsevier, 1965.

6.22. Winslow, F. H. and Hawkins, W. L. Degradation and stabilization. In (Raff, R. A. V. and Doak, K. W., eds.) *Crystalline Olefine Polymers*, Vol. 2, Ch. 8. New York: Wiley-Interscience, 1965.

6.23. Zolotova, N. V. and Denisov, Ye. T. Mechanism of propagation and degenerate chain branching in the oxidation of polypropylene and polyethylene. *J. Polymer Sci.* **A1**(9):3311 (1971).

Chapter 7
Photodegradation

Polymers have different photodegradative sensitivities to UV light of different wavelengths. The varying sensitivities result from differences in the chemical structure. As shown in Table 7.1, the maximum sensitivity of several polymers (as determined by the bond dissociation energies) is in the range of 290 and 400 nm. Although the atmosphere of the earth filters out the UV part of solar radiation, the above (so-called actinic) range of solar ultraviolet radiation is about 6% of the total radiation of the sun which reaches the earth's surface (Figure 7.1). In many applications of plastics, sunlight is a significant source of degradative energy; hence, the study and understanding of the physical and chemical processes caused by light are of great practical importance.

7.1 PHOTOPHYSICAL PROCESSES

The physical processes involved in photodegradation include absorption of light by the material, electronic excitation of the molecules, and deactivation by radiative or radiationless energy transitions, or by energy transfer to some acceptor. When the lifetime of the excited state is sufficiently long, the species can participate in various chemical transformations.

Phenomenologically, the *absorption of light* can be described by Beer-Lambert's law. The intensity of the incident radiation (I_0) will be only partly transmitted (I) through the material; the logarithm of the transmittance ($T = I/I_0$), is proportional to the thickness of the layer (l) and to the concentration of the absorbing component (c):

$$A = -\log T = \log \frac{I_0}{I} = \epsilon l c \qquad (7.1)$$

137

Table 7.1. Wavelength of UV Radiation (Energy of a Photon) at Which Various Polymers Have Maximum Sensitivity.

POLYMER	NM	KCAL/MOL
Styrene-acrylonitrile copolymer	290, 325	99, 88
Polycarbonate	295, 345	97, 83
Polyethylene	300	96
Polystyrene	318	90
Polyvinyl chloride	320	89
Polyester	325	88
Vinyl chloride–vinyl acetate copolymer	327, 364	87, 79
Polypropylene	370	77

where the proportionality factor is the absorption coefficient (ϵ). A (also known as the extinction, E) is dimensionless; i.e., when l is given in cm and c in mol/liter, then the dimension of ϵ is liter mol^{-1} cm^{-1} (molar absorption – or extinction – coefficient).

The absorption of light results in an electronic transition between two energy levels in the absorbing molecule; this absorbed energy is exactly equal to the energy of a light quantum:

$$\Delta E = h\nu \tag{7.2}$$

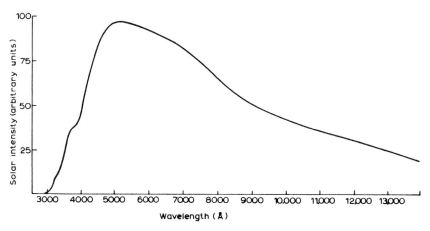

Figure 7.1. UV and visible spectrum of sunlight at noon in midsummer in Washington, D.C. (Reprinted, by permission of Elsevier Publishing Company, from reference 7.5.)

where h is Planck's constant and ν is the frequency of the absorbed light:

$$\nu = \frac{c^*}{\lambda} = c^*\nu^* \tag{7.3}$$

where c^* is the velocity, λ is the wavelength, and ν^* is the wave number of the absorbed light (a possible set of values and dimensions: $h = 6.62 \times 10^{-27}$ erg sec, ν sec^{-1}, $c^* = 3 \times 10^{10}$ cm sec^{-1}, λ cm, ν^* cm^{-1}).

The energy absorption produces an *excited state* of the molecule, two types of which can be distinguished. In a singlet state (S) the spins of the electrons are (remain) paired, in a triplet state (T) they are unpaired. The ground state is almost always a singlet state (S_0). An excitation of a molecule from the ground state to the first excited singlet state ($S_0 \to S_1$) is shown in Figure 7.2. Here, the curves represent the potential energies of the corresponding states and the horizontal lines are the various vibrational energy levels. The excited molecule can lose its excess energy by vibrational relaxation and emission (*fluorescence*). *Radiationless transitions* (internal conversion) from higher excited singlet states (S_2, S_3, etc.) to the S_1 state, or from S_1 to the S_0 state are also possible.

A direct excitation from S_0 to an excited triplet state is not allowed. This transition is, however, possible by *intersystem crossing*, an example of which is

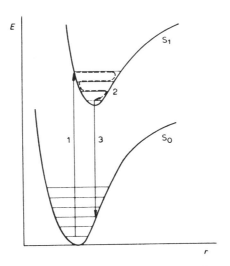

Figure 7.2. Potential energy curves of a molecule (S_0 = ground state, S_1 = first excited singlet state): (1) absorption; (2) vibrational relaxation; (3) fluorescence. (Reprinted, by permission of Elsevier Publishing Company, from reference 7.5.)

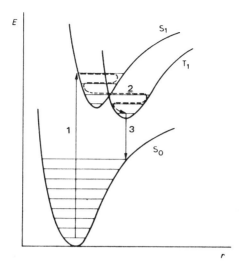

Figure 7.3. Potential energy curves of a molecule (S_0 = ground state, S_1 = first excited singlet state, T_1 = first excited triplet state): (1) absorption; (2) intersystem crossing by vibrational relaxation; (3) phosphorescence. (Reprinted, by permission of Elsevier Publishing Company, from reference 7.5.)

shown in Figure 7.3. The condition of such a transition is that the potential curves of S_1 and T_1 have a common (crossing) point with identical nuclear configuration so that a vibrational transmission ($S_1 \rightarrow T_1$) is possible. The radiative deactivation $T_1 \rightarrow S_0$ is also forbidden; therefore, the T_1 excited triplet state has a much longer lifetime then the S_1 state. The emission of light by $T_1 \rightarrow S_0$ transition is called *phosphorescence.*

When the lifetimes are sufficiently long, a *bimolecular deactivation* of the excited states (deactivation by collision) is also possible. This is very important because it makes the application of deactivators (suitable energy acceptors) possible. This kind of energy transfer is called quenching; the addition of quencher molecules to the polymer is a common method of photostabilization.

There are, however, also other possibilities for energy transfer from an excited species, e.g., *radiative transfer* in which the light emitted during deactivation is absorbed by another molecule, or *nonradiative transfer* occurring over quite large distances (50–100 Å; so-called resonance excitation transfer) or over shorter distances without a real collision but by overlapping of the electron clouds of the participating species (10–15 Å; referred to as exchange energy transfer).

7.2 PHOTOCHEMICAL PROCESSES

The chemical processes of photodegradation include isomerization, dissociation, and decomposition of a molecule as a direct consequence of its photophysical excitation, as well as those nonunimolecular chemical reactions which are facilitated by the absorbed energy. Obviously, a photochemical reaction can take place only during the lifetime of the excited state; such a reaction must compete with the physical modes of deactivation.

Photodissociation occurs when the excitation reaches a point above the dissociation limit of the excited protential curve, as illustrated in Figure 7.4a, or when a dissociative excited state is formed (Figure 7.4b) in which repelling of the atoms occurs at any separation distance. Dissociation can also occur after intersystem crossing to an excited triplet state above the dissociation limit (Figure 7.5a) or to a dissociative potential curve (Figure 7.5b).

An especially important case of *photodecomposition* is the radiation-induced decomposition of *hydroperoxides* formed during polymer oxidation. The energy of UV light is sufficient to cause both of the following decompositions:

$$ROOH \xrightarrow{h\nu} RO\cdot + \cdot OH \quad (\sim 42 \text{ kcal/mol}) \tag{7.4}$$

$$ROOH \xrightarrow{h\nu} R\cdot + \cdot OOH \quad (\sim 70 \text{ kcal/mol}) \tag{7.5}$$

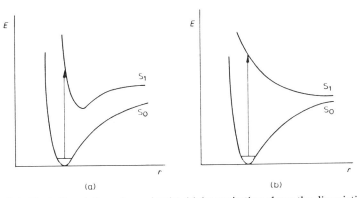

Figure 7.4. Photodissociation of a molecule: (a) by excitation above the dissociation limit of the S_1 potential curve; (b) by formation of a dissociative excited state. (Reprinted, by permission of Elsevier Publishing Company, from reference 7.5.)

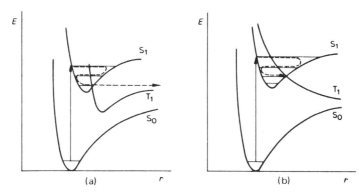

Figure 7.5. Photodissociation of a molecule after intersystem crossing: (a) to above the dissociation limit of the T_1 potential curve; (b) to the dissociative excited triplet state. (Reprinted, by permission of Elsevier Publishing Company, from reference 7.5.)

Dissociation of the O—H bond is less facile:

$$ROOH \xrightarrow{h\nu} ROO\cdot + \cdot H \quad (\sim 90 \text{ kcal/mol}) \tag{7.6}$$

Because of low bond dissociation energy, decomposition according to reaction (7.4) is predominant in polymer photooxidation.

Another important decomposition reaction is the *photolytic scission* of the C—Cl bond in PVC:

$$\sim\!\!-CH_2-\underset{\underset{Cl}{|}}{CH}-\!\!\sim \xrightarrow{h\nu} \sim\!\!-CH_2-\underset{\cdot}{CH}-\!\!\sim + \cdot Cl \tag{7.7}$$

which can initiate the photodegration of the polymer. An obvious difference from thermal degradation of PVC (see Chapter 5) is that in photodegradation of PVC (similar to thermo-oxidative degradation) the participation of radicals in the process is more pronounced.

The role of carbonyl groups in polymer photooxidation is of great importance because once they have been formed they absorb UV light readily, hence, excitation to singlet and triplet states is easy. The excited *carbonyl* groups decompose via *Norrish reactions* of types I, II, and III. The Norrish-I reaction is a radical cleavage of the bond between the carbonyl group and the α-carbon atom (α-scission), and is usually followed by the formation of carbon monoxide:

$$\sim\!-CH_2-\overset{\overset{\textstyle O}{\|}}{C}-CH_2-\!\sim\ \xrightarrow{h\nu}\ \sim\!-CH_2-\overset{\overset{\textstyle O}{\|}}{C}\cdot\ +\ \cdot CH_2-\!\sim \qquad (7.8)$$

$$\downarrow$$

$$\sim\!-CH_2\cdot\ +\ CO$$

The Norrish-II reaction is a nonradical scission which occurs through the formation of a six-membered cyclic intermediate. Abstraction of a hydrogen from the γ-carbon atom results in decomposition by β-scission to an olefin and an alcohol or ketone. For example, in the case of polyethylene, a terminal double bond and an enol/keton end group are formed:

$$(7.9)$$

The Norrish-III reaction is also a nonradical chain scission; however, it involves the transfer of a β-hydrogen atom and leads to the formation of an olefin and an aldehyde:

$$\sim\!-\overset{\overset{\textstyle O}{\|}}{C}-\overset{\overset{\textstyle CH_3}{|}}{CH}-\!\sim\ \xrightarrow{h\nu}\ \sim\!-\overset{\overset{\textstyle O}{\|}}{CH}\ +\ \overset{\overset{\textstyle CH_2}{\|}}{CH}-\!\sim \qquad (7.10)$$

The activation energies of the Norrish reactions are different; the probability of Norrish-II ($E_a = 0.85$ kcal/mol) is higher at room temperature than that of Norrish-I ($E_a = 4.8$ kcal/mol); the latter is, however, more probable at higher temperatures.

7.3 PHOTOOXIDATION

Polymer photooxidation is very similar to thermal oxidation of polymers; its mechanism includes the same reactions shown in Figures 6.1, 6.4, and 6.5. Significant differences exist with respect to the photochemical decomposition of the hydroperoxide and carbonyl groups, as well as with regard to the *initiation* reaction. The formation of polymer radicals

$$RH \xrightarrow{h\nu} R\cdot + H\cdot \qquad (7.11)$$

by scission of a C—H bond is a possible consequence of UV irradiation. The probability of reaction (7.11) is higher than that of the direct reaction between molecular oxygen and a polymer [see (I.1) in Figure 6.1], although the probability of the latter reaction may be increased due to UV excitation.

Not only can the polymer be in an excited state, but the oxygen as well. There are two types of excited *singlet oxygen* ($^1O_2$) having different excitation energies above their ground state. The lower (more stable) excited singlet state ($^1\Delta_g$) has an energy excess of 22.5 kcal/mol; the higher ($^1\Sigma_g^+$), 37.5 kcal/mol. Singlet oxygen can be formed by direct irradiation of O_2; although this excitation is unfavorable, it is possible in the upper layers of the atmosphere. Ozone photolysis, which is very rapid in the upper atmosphere, leads to the formation of singlet oxygen. Singlet oxygen also forms easily in polluted (urban) atmospheres. This may cause the rapid deterioration of polymers in urban areas. Molecular oxygen can act as a deactivator of various excited species present in the polymer. Quenching usually results in formation of singlet oxygen. The lifetime of singlet oxygen in solution strongly depends on the solvent; for example, the lifetime of $^1O_2$ ($^1\Delta_g$) in methanol is 2 μsec, in benzene 24 μsec, and in carbon tetrachloride 700 μsec. Besides initiation of oxidation by hydrogen abstraction from a saturated hydrocarbon

$$RH + {}^1O_2 \, ({}^1\Delta_g) \rightarrow R\cdot + HO_2\cdot \qquad (7.12)$$

initiation may occur by the addition of singlet oxygen to unsaturated bonds present in polymers.

Another significant difference between thermal and photooxidation is that many of the oxidation products incorporated in the polymer absorb UV light better than the original polymer. Thus the autoaccelerative character of the process is even more pronounced in photooxidation, and relatively high oxidation rates can be measured at much lower temperatures in the presence than in the absence of light. Completely "pure" polymers do not exist. During synthesis, processing, and storage, various amounts of carbonyl and hydroperoxy groups accumulate in the material. These chromophores absorb UV light to a much greater extent than the polymer. Initiation of degradation consists mainly of the decomposition of these chromophores. Thus, because of the high initiation rate and because of the short kinetic chains resulting from the lower temperatures, the autocatalytic character of photooxidation is usually hidden. Figure 7.6 compares the oxygen absorption of linear polyethylene samples at 100°C in the dark and at 30°C when exposed to UV radiation. (Notice that the time axis is scaled in hours; compare with Figure 6.11.)

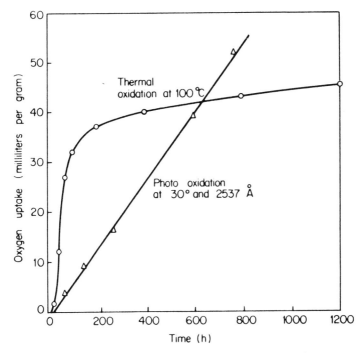

Figure 7.6. Oxygen absorption of linear polyethylene samples at 100°C without, and at 30°C with, UV irradiation. [From F. H. Winslow, W. Matreyek and S. M. Stills, *Polymer Preprints* 7:390 (1966).]

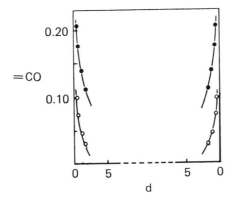

Figure 7.7. Relative amount of carbonyl ($>CO$) groups formed as a function of the distance (*d*) from the surface of a 22 μm polypropylene film. Irradiation time; $\circ$ = 40 hours; $\bullet$ = 80 hours. (From reference 7.3. Copyright 1970. Reprinted by permission of John Wiley & Sons, Ltd.)

As previously shown, sample thickness is an important factor in thermal oxidation owing to its governing of transport processes (O_2 diffusion into the polymer, effusion of volatile products). In photooxidation the sample thickness is even more important than in thermal degradation. As shown in Figure 7.7, the extent of oxidation decreases exponentially with increasing distance from the surface: oxidation mainly occurs in a very thin surface layer of the material.

In polypropylene short hydroperoxide sequences, similar to those formed in thermal oxidation [see reaction (6.1)], are formed during photooxidation, due to intramolecular propagation. Photolysis of these hydroperoxides to yield $\cdot$ OH and $\cdot CH_3$ radicals is responsible for the large amount water and CH_4 formed:

$$\sim\!\!-CH_2-\underset{\underset{CH_3}{|}}{\overset{\overset{OOH}{|}}{C}}-CH_2-\!\!\sim \overset{h\nu}{\longrightarrow} \sim\!\!-CH_2-\underset{\underset{CH_3}{|}}{\overset{\overset{O\cdot}{|}}{C}}-CH_2-\!\!\sim + \cdot OH \nearrow^{H_2O} \tag{7.13}$$

$$\downarrow$$

$$\sim\!\!-CH_2-\overset{\overset{O}{\|}}{C}-CH_2-\!\!\sim + \cdot CH_3 \nearrow^{CH_4}$$

Norrish-I reactions of the formed ketone sequence, followed by H-abstractions, lead to the formation of acetone and carbon monoxide:

$$\sim\!\!-CH_2-\overset{\overset{O}{\|}}{C}-CH_2-\overset{\overset{O}{\|}}{C}-CH_2-\overset{\overset{O}{\|}}{C}-CH_2-\!\!\sim \longrightarrow\!\!\longrightarrow\!\!\longrightarrow$$

$$\sim\!\!-CH_2\cdot + \underline{CO} + \underline{CH_3-\overset{\overset{O}{\|}}{C}-CH_3} + \underline{CO} + \cdot CH_2-\!\!\sim \tag{7.14}$$

This reaction is not important in polyethylene because the polymer contains only isolated hydroperoxide (and carbonyl) groups.

The molecular weight of the polyolefins rapidly decreases during photooxidation. The average number of chain scissions (s) in a polypropylene sample as a function of irradiation time is shown in Figure 7.8. After longer UV irradiation, microcracks appear and the sample becomes opaque and brittle. In stabilized samples, brittleness sometimes occurs before oxygen containing groups appear in the polymer.

Photodegradation at low temperatures is favorable to random scission processes. Polymethyl methacrylate and poly-α-methylstyrene, known to depolymerize quite readily, have only a low monomer yield when irradiated at room temperature, due to the low kinetic chain length. For example, the maximum

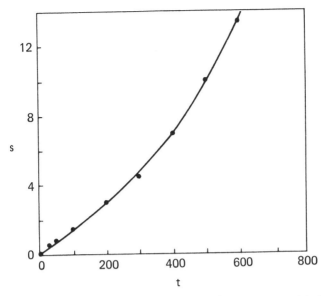

Figure 7.8. Average number of chain scissions ($s = P_{n,0}/P_n - 1$) in one original macromolecule as a function of time (t, in hours) during UV irradiation of a polypropylene sample. (From reference 7.1. Copyright 1970. Reprinted by permission of John Wiley & Sons, Ltd.)

zip length in PMMA is about five monomer units per absorbed quantum; however, at 160°C this value is as high as 220. The mechanism of polyacrylonitrile photodegradation is also different from that of the thermal process. At low temperatures, UV irradiation leading to random scission dominates and side chain polymerization, characteristic of thermal PAN degradation, is negligible.

As previously mentioned (this will be discussed in connection with sensitized photodegradation), various chromophores originally present in the polymer, accumulated during processing and storage, or added intentionally to the polymer may have an important effect on photostability. Metallic impurities, e.g., residues of Ziegler-Natta catalysts or metallic traces originating from processing equipment, may have a catastrophic effect because they can participate in various steps of photooxidation. Metal compounds can, however, be useful in some cases, e.g., by decomposing hydroperoxides without formation of radicals or by functioning as UV absorbers. These possibilities will be discussed in connection with photostabilization of polymers.

7.4 SENSITIZED PHOTODEGRADATION

The ever-increasing amount of plastics used for disposable packaging material and their potential for causing permanent pollution of the environment have prompted workers to seek methods of producing polymers with *controlled service life*. After having finished its useful function, e.g., as a container for milk or as a sandwich bag, the plastic becomes a waste material; it is desirable that this waste decompose with the aid of sunlight, humidity, and bacteria as rapidly as possible. Another type of application in which controlled lifetime is a significant feature of the material is the use of plastic mulch films for protection of important plants by covering the soil (increasing temperature, retaining soil humidity, suppressing weeds, etc.). When the plant has grown and strengthened, the film should disappear, again partly with the aid of sunlight. In agricultural and horticultural applications, however, the material must spend its useful life outdoors. Mulch films must be sensitized for photodegradation but also stabilized to an extent which insures their preplanned service life. After the consumption of stabilizer, the remainder of photosensitizer causes rapid degradation of the polymer to low molecular weight compounds which are small enough to be decomposed by microorganisms. Thus, the complete deterioration of plastic wastes and the control of service life of plastics are rather complex procedures.

A photosensitizer usually has a high absorption coefficient for UV light; the excited compound either decomposes into free radicals and initiates degradation or oxidation of the polymer, or it transfers the excitation energy to the polymer (or to oxygen). A good sensitizer should be easily admixed with the polymer and must not decompose thermally or in the dark.

Polycyclic aromatic compounds, e.g., naphthalene, anthracene, pyrene, etc., are assumed to generate singlet oxygen by energy transfer to ground state oxygen molecules. *Aromatic ketones* and diketones abstract hydrogen or decompose after excitation and cause the initiation of photochemical processes. For example, excited benzophenone can abstract hydrogen from the polymer, or the radical formed from benzophenone can react with oxygen and form a $HO_2 \cdot$ radical and regenerate a benzophenone molecule:

$$\tag{7.15}$$

Quinones behave similarly: their excited state has biradical character, the biradical abstracts hydrogen from the polymer and hydroquinone forms. For example, the addition of various quinones sensitizes the photodegradation of polyisoprene solutions. The sensitizing effect is drastic, especially with anthraquinone.

Various *nitrogen containing chromophores*, e.g., azo, nitroso, and aromatic amino compounds, decompose to radicals when excited by UV light. For example, 2-chloro-2-nitroso-propane was found to sensitize photodegradation of polyisoprene even more effectively than anthraquinone. Some N-halides are also active sensitizers, e.g., trichlorosuccinimide. *Organic disulfides*, like peroxides, photodecompose to form radicals, hence causing initiation of the degradation process. *Dyes* can also be applied for sensitizing polymer photodegradation. For example, the photodegradation of *cis*-1,4-polyisoprene can be significantly accelerated by the addition of a small amount of methylene blue.

Both organic and inorganic metal compounds can sensitize photodegradation. Especially interesting is the fact that certain *iron complexes*, e.g., Fe(II) and Fe(III) complexes of dithiocarbamates and 2-hydroxyacetophenone oximes, exhibit a delayed action as UV sensitizers. Fe(III) acetylacetonate (A) has an immediate effect; however, the effect of the Fe(III) complex of 2-hydroxy-4-methylacetophenone (B), as shown in Figure 7.9, can be observed only after an induction period of significant length. Ferrocene derivatives such as benzoyl-ferrocene (C) can act as photosensitizers.

(7.16)

Inorganic metal oxides and salts, e.g., ZnO, TiO_2, and $FeCl_3$ accelerate photodegradation. It is assumed that free radicals are formed from these compounds during irradiation.

Another type of photosensitized polymer contains the sensitizer in the form of chromophores chemically bound to the polymer or incorporated as comonomer units into its backbone. Such *photodegradable polymers* have been prepared by copolymerization of ethylene with carbon monoxide, of methyl methacrylate with methyl vinyl ketone, and of styrene with phenyl vinyl ketone. Among the copolymers containing carbonyl groups, the yield of photolytic scission is small when CO is in the backbone and relatively high when CO is in a pendant group. Vinyl polymers and polyamides have been prepared with pyridine or pyrazine

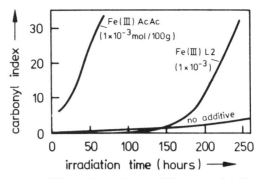

Figure 7.9. Carbonyl index (IR absorbance due to $\geq$CO groups related to a reference absorbance) as a function of time during UV irradiation of low density polyethylene samples: Fe(III)AcAc = Fe(III) acetylacetonate; Fe(III)L2 = Fe(III) 2-hydroxy-4-methylacetophenone. (Reprinted, by permission of Elsevier Publishing Company, from reference 7.12.)

rings in the backbone. Polymers with unsaturated bonds in the main chain are sensitive to photodegradation.

REFERENCES

7.1. Adams, J. H. Analysis of nonvolatile oxidation products of polypropylene. III. Photodegradation, *J. Polymer Sci.* **A1**(8):1279 (1970).

7.2. Birks, J. B. *Photophysics of Aromatic Molecules.* New York: Wiley-Interscience, 1970.

7.3. Carlsson, D. J. and Wiles, D. M. Surface changes during the photo-oxidation of polypropylene. *J. Polymer Sci.* **B8**:419 (1970).

7.4. Carlsson, D. J. and Wiles, D. M. The photo-oxidative degradation of polypropylene. *J. Macromol. Sci., Reviews Macromol. Chem.* **C14**:65, 155 (1976).

7.5. Geuskens, G. Photodegradation of polymers. In (Bamford, C. H. and Tipper, C. F. H., eds.) *Comprehensive Chemical Kinetics*, Vol. 14. Amsterdam: Elsevier, 1975.

7.6. Geuskens, G. and David, C. The photo-oxidation of polystyrene. In (Geuskens, G., ed.) *Degradation and Stabilization of Polymers.* London: Applied Science Publishers, 1975.

7.7. Guillet, J. E. and Norrish, R. G. W. The photolysis of polymethylvinylketone. I. Reactions and kinetics. *Proc. Royal Soc., Ser. A* **233**:153 (1956).

7.8. Guillet, J. E. Fundamental processes in the UV degradation and stabilization of polymers. *Pure and Applied Chemistry* **30**:135 (1972).

7.9. Margolin, A. L., Postnikov, L. M. and Shlyapintokh, V. Ya. The mechanism of photochemical aging of aliphatic polyamides. *Polymer Science USSR* **19**:2236 (1977).

7.10. Rånby, R. and Rabek, J. F. *Photodegradation, Photo-Oxidation and Photostabilization of Polymers.* New York: Wiley, 1975.

7.11. Rohatgi-Mukherjee, K. K. *Fundamentals of Photochemistry.* New York: Wiley, 1978.

7.12. Schnabel, W. and Kiwi, J. Photodegradation. In (Jellinek, H. H. G. ed.) *Aspects of Degradation and Stabilization of Polymers.* Amsterdam: Elsevier, 1978.
7.13. Trozzolo, Λ. M. and Winslow, F. H. Λ mechanism for the oxidative photodegradation of polyethylene. *Macromolecules* 1:98 (1968).
7.14. Winslow, F. H., Matreyek, W. and Stills, S. M. Contrasts in thermal and photo-oxidation of polyethylene. *Polymer Preprints* 7:390 (1966).

Chapter 8
Biodegradation

Biodegradation or biological degradation of polymers consists of those processes which result from an attack on the material by living organisms, e.g., bacteria, fungi, insects, and rodents. The term is, however, usually restricted to the degradation caused by microorganisms; a more general term which includes all kind of attacks by living creatures is sometimes called biodeterioration.

The study of biodegradation has two opposite facets. On the one hand, there are many applications where *biologically resistant* materials are needed. In such applications, the polymer is exposed to attack by various microorganisms and must resist them as long as possible. Telecommunication and high tension cables are often buried or laid under-water. Dental, orthopedic, and other surgical implants are exposed to biological attack in the human body. Insulation, paints, and coatings are subject to environmental deterioration including attacks by microorganisms. In all of these applications, the polymer is expected to serve for a long time: it must be bioresistant. Fortunately, most high molecular weight synthetic polymers fulfill this requirement, and the problem is usually restricted to the selection of compounding additives which are also satisfactorily bioresistant or are of fungicidal or bactericidal nature.

On the other hand, there is an increasing demand for *biodegradable* plastics. Large amounts of polymers are used in packaging applications and will be discarded after use. Environmental protection requires self-destructing polymers which will degrade and disappear completely when subjected to the combined effects of various environmental factors (including microorganisms). Biodegradable polymers also are needed for some agricultural and horticultural applications. Mulches used for the protection of sprouting plants must be destroyed after their useful life of several weeks. Biological decomposition of polymers in a controlled manner is necessary in some medical applications, e.g., in surgical sutures and in controlled-release drugs.

Biodegradation is fundamentally a chemical process involving *enzymes* produced by microorganisms. High humidity, moderately increased temperature, and darkness are usually favorable for proper functioning of microorganisms. The optimal conditions, especially with respect to pH and oxygen requirements, are different for various fungi and bacteria. Aerobic bacteria require oxygen for intense growth and enzyme production. The resistance of a polymer to biological attack depends on the availability of specific sites in it for the reaction and enzyme in question. Thus, for enzymatic hydrolytic scission of a polymeric main chain, the presence of hydrolyzable linkages such as amide, ester, or urethane is required.

8.1 METHODS OF TESTING BIODEGRADABILITY

There are three main groups of procedures for studying biological decomposition of plastic materials and their additives. Closest to the actual conditions is the *soil burial method*: standard samples of the material are buried in various soils, in outdoor locations or under laboratory conditions. From time to time, samples are withdrawn and analyzed for signs and quantitative characteristics of decomposition, such as weight loss. This method lacks reproducibility because of the very different microbiological composition and organic material content of soils of various origin (from garden soil to sewage sludge) and the difficulties in controlling climatic factors (temperature and humidity).

More reproducible results can be obtained by using *cultured fungi or bacteria.* The material to be investigated serves as the nutrient medium (alone or combined with agar), and the growth rate of the inoculated fungi (e.g., *Aspergillus niger*) or bacteria (e.g., *Pseudomonas aeruginosa*) is measured after incubation for a convenient period of time. ASTM recommendations describe these methods; they distinguish among five colony growth categories: "0" = no visible growth, "1" = less than 10%, "2" = 10-30%, "3" = 30-60%, and "4" = 60-100% of the surface is covered. Fungi are used more frequently than bacteria.

Besides colony growth, other effects of microbes can also be observed, such as visible destruction of the samples. For quantitative determinations, both the biological oxygen uptake and the production of carbon dioxide can be measured. Comparison of these results with those of control samples renders the application of these methods possible in the case of indoor (laboratory) soil burial. Radioactively labeled polymer can be used, and the evolution of ^{14}C-labeled CO_2 can be detected. The measurement of the formation of other products, as well as the change of the sample weight and of physical characteristics such as molecular weight, tensile strength, etc., can also be used to assess the extent of biological degradation.

As already mentioned, biodegradation proceeds in many cases via chemical reactions catalyzed by enzymes which are synthesized by microorganisms and have very specific action. Testing of biodegradability can be accelerated by using *isolated and purified enzymes* instead of the microbes themselves. Thus the study of the details of degradation is possible, and reaction conditions such as temperature, pH, pressure, and irradiation may be easily varied.

8.2 BIODEGRADABILITY OF POLYMERS

Synthetic polymers are generally resistant to biological attacks. Microorganisms have not yet had time enough for adaptation and synthesis of polymer-specific enzymes. Low molecular weight, the presence of certain kinds of enchainments or end groups, or an increase in polarity and hydrophilicity may increase susceptibility to biodegradation.

Polystyrene and its copolymers are very stable, even when pyrolyzed or photodegraded to low molecular weight fragments. An exception is the copolymer of styrene with vinyl ketone (Ecolyte PS) which is biodegradable, although not as much as other copolymers of vinyl ketones (Ecolyte EE with ethylene or Ecolyte PP with propylene).

Linear, high molecular weight polyethylene is very inert. Photolytic, oxidative, or thermal fragmentation, leading to products with molecular weights lower than 1000, results in a biodegradable material which, like normal low molecular weight paraffins, can readily be utilized by microorganisms. Chain branching decreases susceptibility to attack; branched hydrocarbons are more bioresistant than their linear isomers.

Polypropylene and other high molecular weight polyolefins are reported to be bioresistant materials. The same is true of polyacrylonitrile and polyvinyl chloride; the latter, however, is nonbiodegradable only when used without additives. Polyvinyl acetate is somewhat less bioresistant than other vinyl polymers.

Natural polymers are generally readily attacked by microorganisms. Most of these compounds contain hydrolyzable linkages. The enzymatic hydrolysis of main chain linkages is the main course of biological attack. Thus, cellulose, starch (and other polysaccharides), natural rubber, and proteins are easily degraded by various fungi and bacteria. It should be noted, however, that hydrolysis, oxidation, etc., of these polymers may also occur without the presence of microorganisms, although the rate factor (acceleration by enzymes) is very important.

Many synthetic polymers containing hydrolyzable linkages, such as polyesters, polyamides, and polyurethanes, are prone to biodegradation. Aliphatic polyesters are more sensitive than aromatic: e.g., poly-ϵ-caprolactone is one of the few commercially available high molecular weight biodegradable polymers; it belongs to the ASTM "4" colony growth category and achieves nearly 100% weight loss in one year when tested by soil burial. Polyamides of high molecular weight are quite resistant to microbial attack; low molecular weight cyclic and linear oligomers of ϵ-caprolactone have been reported to be vulnerable to bacterial attack. The introduction of amino acid groups (e.g., glycine) into the polyamide chain increases the biodegradability. Polyester-base polyurethanes are more sensitive than polyether-linked ones. It has been reported that the higher the conformational flexibility of the chain, the easier is the attack by enzymes on the hydrolyzable linkages.

8.3 BIODEGRADATION OF COMMONLY USED ADDITIVES

Polymer compounds for practical uses are generally made by including additives such as plasticizers, lubricants, and stabilizers in the formulations. In general, these additives have a great influence on the degradability of the final product, in both a positive and a negative respect. The role of additives is especially important, however, with respect to biodegradability. Low molecular weight additives, usually containing various reactive groups, are much more sensitive to microbial attack than the polymer molecules. It has been shown that many of the customary additives belong to the most sensitive colony growth category. Among the additives used for PVC compounding, epoxidized soybean oil (plasticizer), dibutyltin dilaurate (heat stabilizer), and zinc stearate (lubricant) are rated "4" when tested by the ASTM method.

On one hand, biodegradation of additives results in loss of the effect for which the compound was added (e.g., heat stabilization). On the other hand, the biodegradable additive functions as a nutrient, "luring" the microorganisms into the plastic and serving as a good medium for their growth, thus promoting their attack on a polymer molecule otherwise "unattractive" to them. In the case of PVC, the removal of additives by microorganisms and the enhanced susceptibility of the polymer to biological and other degradation lead to an increase in the tensile modulus and to stiffness, increased rigidity, and brittleness of the material.

8.4 BIODEGRADABLE POLYMERS

Accordingly, there are two possible ways to obtain biodegradable plastic materials:

1. Fragmentation of the polymer by sensitized photodegradation may occur (see Chapter 7.4). If the polymer is attractive enough for the microorganisms (e.g., linear polyethylene), microbial attack begins after the polymer had degraded to a resonably low molecular weight. Photodegradation alone is not a sufficient solution: the resulting debris or powdery material is still a contaminant if it cannot be consumed by plants, thus returning to the natural (biological) cycle.

2. Directly biodegradable materials may be produced either by the introduction into the polymer chain, via copolymerization, of specific groups which are sensitive to microbial attack or by compounding the polymer with biodegradable additives. Polycaprolactones and vinyl ketone copolymers have already been mentioned as commercially available biodegradable polymers. A biodegradable polyethylene has been produced by using starch as a filler.

Interest in the study of biodegradable plastic materials is increasing. As of now, very few practical solutions exist, and these cannot meet all the complex requirements of this field.

REFERENCES

8.1. Coscarelli, W. Biological stability of polymers. In (Hawkins, W. L., ed.) *Polymer Stabilization.* New York: Wiley-Interscience, 1972.
8.2. Guillet, J. E. Polymers with controlled lifetimes. In *Proceedings on the Degradability of Polymers and Plastics.* London: The Plastics Institute, 1973.
8.3. Guillet, J. E. (ed.). *Polymers and Ecological Problems.* New York: Plenum Press, 1973.
8.4. Nykvist, N. P. Biodegradation of low density polyetylene. In *Proceedings on the Degradability of Polymers and Plastics.* London: The Plastics Institute, 1973.
8.5. Potts, J. E., Clendinning, R. A., Ackart, W. P. and Niegisch, W. D. The biodegradability of synthetic polymers. *Polymer Preprints* 13:629 (1972).
8.6. Potts, J. E. Biodegradation. In (Jellinek, H. H. G., ed.) *Aspects of Degradation and Stabilization of Polymers.* Amsterdam: Elsevier, 1978.

Chapter 9
Mechanical Degradation

Mechanical degradation of a polymer includes, in a broad sense, every kind of mechanically induced breakdown of the material. Thus, irreversible deformation, crazing, cracking, fracture, and fatigue of polymeric bodies under static or dynamic loading are processes which lead to mechanical deterioration. These types of material failure are usually circumstantially treated in texts on the *mechanical properties* of polymers, especially regarding their phenomenological, physical, and engineering aspects. In a narrower sense, the molecular understanding and description (i.e., the chemistry) of the mechanically (stress) induced processes can be considered as the subject of studies on the *mechanical degradation* of polymers.

The term *mechanochemistry* covers a broader field: it deals with those processes which utilize stress-induced chemical changes to transform polymers into new materials (e.g., synthesis of block and graft copolymers). Thus mechanochemistry includes both mechanical degradation and synthesis. *Chemomechanics* is the word suggested for mechanical changes (alteration of the mechanical properties) caused by chemical reactions (e.g., oxidation of the polymer) or by various environmental factors.

9.1 METHODS OF INVESTIGATION

Methods for inducing deformation or fracture of a polymer by mechanical forces depend on the physical state and the mechanical properties of the material. In the case of *solid polymeric bodies*, we can study the effect of the applied stress using constant load (e.g., uniaxial load in a tensile test), applying a single blow (impact tests), or submitting the material to intermittent or cyclic loading (dynamic fatigue). Friction, abrasion, and wear and tear are also accompanied by mechanochemical processes which are very important — particularly with respect to the tire industry. Depending on the rigid or elastic nature of the material, different techniques can be applied to produce mechanical degradation. Rigid materials undergoing brittle or ductile (tough) failure (with or without permanent deformation before fracture) can be crushed in vibromills which combine the intense action of impact, compression, and shear. Figure 9.1 shows the schematic sketch of a ball mill apparatus in which crushing (comminution or grinding) can be carried out under vacuum; the apparatus is suitable for taking samples for ESR measurements during milling. Another typical vibromill is shown in Figure 9.2; body and balls are usually made of steel or porcelain. Grinding under various gases or in liquid media renders the study of the combined effect of stress and environment or specific chemical reagents possible. Machining of plastics, e.g., drilling, also produces broken bonds and free radicals. A method for producing new surfaces from nonrigid, solid polymers is slicing, for example, with a razor blade driven by a small motor. Cutting of polymer samples by a steel saw in liquid nitrogen and ESR investigation of the sawdust have also been applied to the study of mechanical degradation.

For the *rubbery state*, mastication is a very old and widely applied technological process used to produce mechanical degradation and subsequent alterations. This technique involves kneading, shearing, and compressing the material, cold or hot, usually in closed mixing (roll) mills, internal mixers, etc. In the scroll masticator (Figure 9.3), the material is loaded between two scrolls having

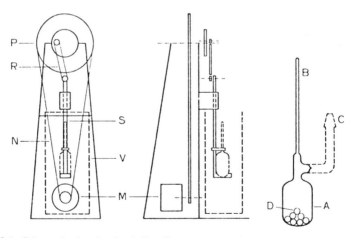

Figure 9.1. Schematic sketch of a ball mill apparatus: (A) glass ampule, (B) ESR sample tube, (C) connector to a vacuum system (sealed off when evacuated), (D) glass balls for milling, (S) holder for ampule, (M) motor, (N) belt, (P) pulley, (R) crank, (V) Dewar flask containing coolant. (Reprinted, by permission, from reference 9.16.)

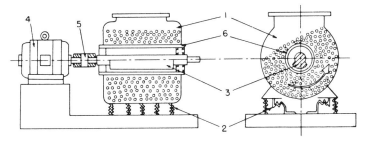

Figure 9.2. Schematic diagram of a vibromill: (1) cylindrical body, (2) elastically sprung base, (3) out-of-balance vibrator, (4) electric motor, (5) flexible coupling, (6) double walls in which cooling water is circulated. (Reprinted, by permission, from reference 9.6.)

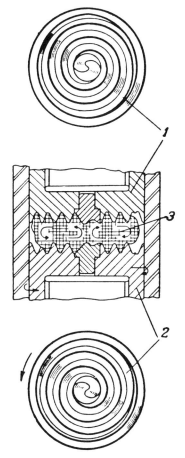

Figure 9.3. Scheme of scroll mastication: (1) top scroll face, (2) bottom scroll face, (3) rubbery material. (Reprinted, by permission, from reference 9.6.)

159

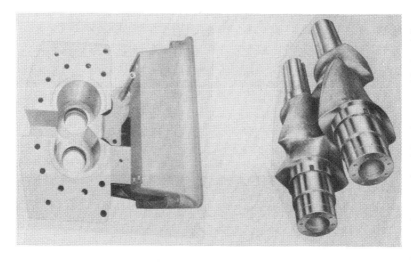

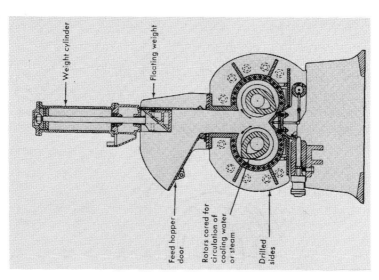

Figure 9.4. Cross section of a Banbury mixer with drilled sides; photographs of the chamber with closed drop door and of the four-wing rotors. (With kind permission of the Farrel Company.)

engraved spirals of convenient design. The top scroll is stationary while the lower rotates at a speed of 75–300 rpm. The rubber is circulated between the scrolls in a complex way and the shear deformation causes its degradation. Banbury mixers (Figure 9.4) are widely used for mastication of rubbers and other elastomers. Their main part is a closed mixing chamber in which the rubbery material is kneaded by two conveniently shaped rotors; the wall of the chamber also participates in the shearing action. The chamber and the rotors can be thermostated. Banbury mixers are also available in laboratory sizes. Torque rheometers, for example, the Brabender Plasticorder, function similarly; they give, however, a time record of the torque change connected with the progress of mechanical degradation during mastication. (An application of torque rheometers for dynamic testing of PVC stability is described in Chapter 5.)

In the *molten state*, because of the high temperatures applied it is usually difficult to separate thermal and mechanical degradation. Some methods of thermomechanical analysis have been mentioned in section 2.1. Screw and plunger extruders, injection molding machines, and internal mixers can be used to investigate shear-induced changes of polymer melts. Application of repeated extrusion causes shear effects to accumulate.

Mechanical degradation of *polymer solutions* can be studied by using viscometric techniques. By forcing a solution through a capillary or by using a rotational viscometer at high shear rate and/or stress, degradation (e.g., molecular weight decrease) can be achieved. Intense shaking, high speed stirring, or turbulent flow may have a similar degradative effect. The McKee Consistometer (and its improved versions) allow repeated passage of a solution through a capillary. High speed stirring instruments (homogenizers), rotating at 10,000–45,000 rpm, are widely used for degradation studies of polymer solutions; however, in addition to shearing forces, other factors (e.g., concentration increase resulting from solvent evaporation) can influence the results. A suitable method for studying mechanical degradation of polymer solutions and dispersions is the use of ultrasonic irradiation; the acoustic energy input, intensity, and frequency of the radiation can be accurately measured. Ultrasonic generators, usually employing quartz or other piezoelectric crystals as oscillators, are commercially available. Freezing of solutions can cause degradation of the dissolved polymer due to the mechanical action of the forming solvent crystals. Repeated freezing and thawing may result in a severe molecular weight decrease. These experiments can be carried out in very simple cooling-heating devices; however, the cooling rate has a considerable effect on degradation.

During mechanical degradation, heat is generated which may increase the temperature of the material. Therefore, the results of mechanical degradation experiments must be carefully analyzed with respect to a possible warming of the system. In some cases, the overlapping of the thermal and mechanical effects can be resolved by using additional investigative methods; the analysis of products, which sometimes differ depending on the route of degradation, may provide complementary information. Another possible way of separating thermal and mechanical effects is to study the temperature dependence of the degradation. An increase of degradation rate with decreasing temperature clearly indicates a mechanical mechanism. For example, the shearing forces decrease with decreasing viscosity when the temperature is raised in mastication experiments; the mastication efficiency decreases with increasing temperature. However, other processes such as autoxidation when oxygen is present may complicate this picture. As shown in Figure 9.5, mastication efficiency changes according to a minimum curve in the presence of oxygen.

An important method of investigating mechanical degradation is electron spin resonance (ESR) spectroscopy (see section 2.3). Free macroradicals are formed when polymer chains rupture. ESR study of degraded samples provides information on the formation and decay of radicals, and facilitates their identification and quantitative determination, thereby contributing to the understanding of the molecular mechanism of mechanical degradation. Another direct investigative technique for the observation and kinetic study of radicals is the addition of

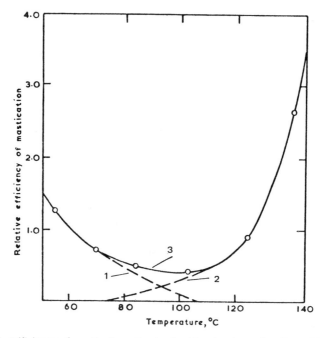

Figure 9.5. Efficiency of mastication of natural rubber in air as a function of temperature: (1) mechanical breakdown, (2) thermo-oxidative breakdown, (3) the overall observed efficiency. (Reprinted, by permission, from reference 9.6.)

free radical acceptors (scavengers) to the degrading system. The consumption of these additives (usually stable free radicals, e.g., diphenylpicrylhydrazyl) by the forming mechano-radicals can be monitored, for example, by colorimetric measurements.

Low angle x-ray scattering and light scattering can be used to obtain information on the formation of microcracks and cavities (microvoids) in stressed material. Infrared (IR) spectroscopy is also a useful, associated technique which detects the bond vibration frequency shifts caused by the applied stress before rupture. IR analysis of the newly formed end groups, on the other hand, provides information about the events after rupture, especially when environmental factors are also involved. In this latter respect, the application of NMR can be very informative. Other techniques which detect mechanical degradation or measure its extent are: the investigation of molecular weight and MWD (viscometry, GPC), surface area measurements, and the study of gel formation. The consequences of mechanical degradation can be detected, of course, by measuring the changes in the mechanical properties of the material.

9.2   MOLECULAR MECHANISM AND THEORIES

In addition to primary chemical bonds in the main polymer chain, polymeric materials contain secondary bonds. Chemical bonds between polymer chains (permanent cross-links), intermolecular van der Waals' bonds, dipole-dipole interactions, segmental attractions and other physical couplings (temporary cross-links), chain entanglements, as well as cross-link density, segment mobility, crystallinity, superstructure, and other morphological features — all contribute to the macroscopic mechanical properties of the material and thus also to its behavior when subjected to mechanical forces. Although chain scission (i.e., breaking of the primary chemical bonds) is the ultimate response of the mechanically stressed material (e.g., when uniaxially loaded in a tensile test), prior to fracture a material may undergo various conformational changes, cavitation, and longitudinal slipping of chains. The stressed chains become mechanically excited. De-excitation can occur by entropy relaxation (conformational change), enthalpy relaxation (cavitation and slippage), or bond scission.

According to the *Peterlin model* of chain rupture in a stretched crystalline polymer (the model was originally constructed to explain rupture characteristics of fibers), there are load-bearing (fully stretched) and nonstretched chains in the material at a given strain (Figure 9.6). During gradual displacement of the crystallites, the shortest tie molecules will be stretched first. After reaching the maximum possible length, these stress-concentrated chains break and mechano-radicals form. The model explains the experimental fact that the radical concentration does not depend on the applied stress but on the strain. However, this

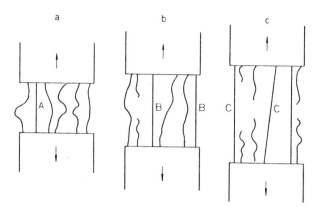

Figure 9.6. Load distribution among tie molecules in the morphous layer of a crystalline polymer (the squares represent crystallites) and their rupture with increasing strain: (a) the fully stretched tie molecule, A, carries all the load; (b) A is broken and two fully stretched tie molecules, B, carry half the load each; (c) A and B are broken and three chains, C, carry one-third of the load each (Peterlin model). (Reprinted, by permission of John Wiley & Sons, from reference 9.12.)

model assumes that the crystalline blocks are sufficiently strong to remain unaffected by the stress field. In amorphous polymers, no crystalline parts — and thus no tie molecules — exist.

Another model of microvoid formation, applicable to both crystalline and amorphous material, is the *Zhurkov model* of submicrocrack generation (Figure 9.7). This model assumes a mechanochemical chain reaction which explains the accelerating effect of stress concentrations at microstructural inhomogeneities. Once a pair of chain end radicals are formed (b) by a bond scission (preferably at a structural defect site), they rapidly transform (by hydrogen abstraction from neighboring chains) to stable end groups, giving rise to the formation of two in-chain radicals (c). These "radicalized" chains easily undergo thermal chain

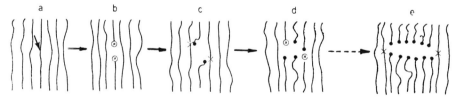

Figure 9.7. Zhurkov model of submicrocrack formation: (a) stretching of a chain; (b) chain scission and formation of chain end radicals ($\otimes$); (c) hydrogen abstraction, formation of stable end groups ($\bullet$) and in-chain radical ($\times$); (d) scission of the radicalized chains; (e) formation of a submicrocrack. (Reprinted, by permission, of the American Institute of Physics, from reference 9.18.)

scission, forming stable end groups and again chain end radicals (d). Repetition of steps c and d (formation of 3000–5000 new end groups per submicrocrack) does not increase the number of free radicals; the number remains two as formed in the initiation step b.

There are several theories which contribute to the understanding of the mechanical failure of polymeric materials. Thus, *Griffith's theory* suggests that the strength of a brittle solid is controlled by flaws (microcracks) already existing on its surface. The propagation (widening) of a preexistent crack, which is a material defect, is governed by the specific material constants (modulus of elasticity, etc.). *Andrews' generalized theory* of fracture mechanics correlates the surface work with the energy necessary to break the bonds in a unit area across the fracture plane ("surface energy"), with a function ("loss function") of the crack propagation rate and with the temperature and the strain applied to the specimen. In perfectly elastic materials, the loss function reduces to unity and the generalized theory gives the same results as Griffith's. The surface energy is determined by the strength of the primary bonds in the fracture area. Although high cross-link density and the presence of tie bonds or entanglements increase the number of primary bonds in the fracture area, the surface energy will not be higher with increasing cross-link concentration. This surprising behavior can be interpreted by the *Lake-Thomas theory* according to which the entire network chain is energized before the weakest bond in the chain breaks. The higher the cross-link density, the smaller is the energy which can be stored in the polymer backbone between cross-links. Thus, as long as secondary bonds (e.g., segmental attractions) do not play a considerable role in restraining crack propagation, increasing cross-link density favors the fracture process.

The stretching, etc., energy applied to an elastic material (e.g., to an elastomer) leads to elastic deformation of the body. However, a significant part of the energy becomes internally dissipated. According to the *theory of Grosch, Harwood, and Payne* the energy required to break an elastomer is proportional to the $\frac{2}{3}$ power of the dissipated energy: the higher the energy dissipated before rupture, the stronger the elastomer. The mechanisms of energy loss within the material may be diverse; however, energy dissipation always increases the resistance to fracture processes.

Plastic deformation, characteristic of glassy polymeric materials, also increases fracture toughness. It has a strong effect on crack propagation. Plastic deformation begins where the stress is the largest, i.e., at the tip of the crack. Thus, plastic deformation increases the radius of the crack tip and renders stresses subcritical. The plastic energy dissipation increases material resistance.

Application of the chemical rate theory to the mechanical degradation of polymers also provides useful results. The change of secondary bonds (e.g., slipping) during uniaxial loading was treated by Tobolsky and Eyring. They predicted a lifetime, the logarithm of which is almost linearly related to the applied stress. The kinetics of primary bond rupture was analyzed (independently) by Zhurkov and Bueche; their lifetime prediction (see section 2.5) has been confirmed in many cases. The Zhurkov and Bueche treatment is based on the assumption that the bond scission is the rate-determing step in the failure process. Macroscopic failure usually results only after the accumulation of the ruptured bonds into cracks. Cracks need time to grow to reach considerable sizes. The random accumulation of bond scissions in the fracture initiation period was computer simulated by Dobrodumov et al. for a regular network model subjected to different loads. As shown in Figure 9.8, in the case of small loads ($\gamma \sigma_0 / kT = 5$) a large number of isolated fracture events happen before crack growth begins. With increasing loads, crack propagation starts at a small number of ruptured bonds. Similar results were obtained by Kausch and Hsiao who studied the effect of network orientation on defect accumulation, using different loads. According to their theory, stiffness has a decisive role in determining material strength because it affects the local stress concentrations; it is presumed that breaking occurs at a critical local strain.

In order to illustrate the role of strain and stress, very interesting experiments were carried out by stepwise straining of a polyamide fiber (nylon 6). The

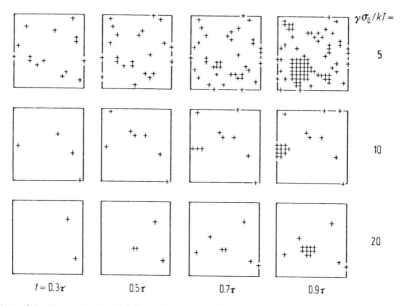

Figure 9.8. Computer simulated random accumulation of bond scissions during the fracture initiation period (τ = time to failure) in a regular network model using different loads (σ_0 = the applied stress): γ, a system constant; k, Boltzmann constant; T, temperature, t, time. (Reprinted, by permission of the American Institute of Physics, from reference 9.8.)

straining leads to sudden stress increase at each step; as a result chain scission (along with mechano-radical formation) begins. During the constant strain period, stress relaxation occurs. Interestingly, the rate of radical formation decreases to practically zero, although the relaxation of the macroscopic tensile stress amounts to only about 10%. The cause of this phenomenon is that the breakable chains have a length distribution and the local stresses are not uniformly distributed among the load-carrying chains. As a result of chain scissions, the stress is more evenly distributed. Mechanical degradation can continue in the next step only when strain and stress are again increased, i.e., when new chains become fully extended (see Figure 9.6) and higher local molecular stresses are produced.

A rather general result of mechanical degradation experiments, especially in investigations of non–solid state samples, is that there is a limiting value of molecular weight (M_∞) toward which a given polymer tends under given conditions of degradation. According to *Frenkel's theory*, in a sheared polymer the molecules are extended in the flow direction with the greatest extension being in their middle part. The chain ends more or less preserve their original shape. The force acting on the individual bonds increases with the square of the distance from the chain end. Under given shear conditions (viscosity, shear rate, etc.), molecules below a critical length are assumed to be stable. Longer chains have higher sensitivity to rupture than shorter ones. (This would be true in a random scission process too, because longer chains contain more bonds.) *Baramboim's theory* suggests that the limiting (critical) molecular weight, i.e., the size of the longest molecule which still can move without chain scission, may be calculated by

$$M_\infty = \frac{ES_2}{qS_1}m \tag{9.1}$$

where E and q, respectively, are the energy of the main chain bonds and of the intermolecular attraction; S_1 and S_2 are the limiting interatomic and intermolecular distances, respectively, at the instant of rupture, and m represents the molecular weight of a monomeric unit. Baramboim describes mechanical degradation by three parameters:

$$\phi_1 = \frac{M_0 - M_t}{M_0 - M_\infty} \tag{9.2}$$

$$\phi_2 = \frac{M_\infty}{M_t} \tag{9.3}$$

$$\phi_3 = \phi_1 \phi_2 \tag{9.4}$$

where M_0 and M_t are the initial and actual (at time t) molecular weights respectively. ϕ_1 is the conversion of degradation; it is in the range 0 to 1. ϕ_3 expresses the relative number of bonds broken at time t. The third parameter, ϕ_2, has been suggested because it does not contain the initial value M_0. The use of M_0 leads to inconveniences in the comparison of polymers with different starting molecular weights. M_0 is sometimes unmeasurable, for instance in the case of insoluble materials.

The Frenkel theory applies to individual molecules and does not take into consideration the role of secondary bonds and entanglements which may be very important in concentrated and viscous systems. According to *Bueche's theory* which has been applied successfully for rubbers, melts, and solutions, macromolecules cannot freely rotate in shear when entanglements are present. Large tensions are produced on the central bonds of the molecules since these bonds, located between two entanglements, are stressed from both sides. These tensions are proportional to the viscosity and shear rate, and also depend on molecular weight. Bueche's theory predicts that shear degradation is rapid for high molecular weight polymers and slow for low molecular weight polymers.

9.3 FACTORS AFFECTING MECHANICAL DEGRADATION

The chemical and physical structures of the polymer have a strong influence on mechanical degradation. The role of the strength of primary bonds in the main chain and the importance of steric factors and resonance stabilization with respect to the further fate of the formed radicals are the same as in other types of chain scission processes (see section 1.3, also Chapters 3 and 4). Intermolecular forces and various attractions between the chains increase the rigidity of the structure leading to an increased tendency toward mechanical degradation. Chemical cross-linking and physical linking (e.g., entanglement and the presence of crystalline regions) which prevent molecular motion (e.g., slipping) of the polymer chains or their segments enhance fracture. Without such molecular "anchorages", polymers are able to undergo large elastic or irreversible deformations; they can yield and flow under stress due to the uncoiling and straightening of the chains. Extensive, distributed fracture is possible only in systems containing a sufficient amount of structural constraints. In addition to *chemical composition*, the initial *molecular weight* (and molecular weight distribution), and the type and amount of *additives* such as lubricants, plasticizers, and reinforcing fillers, are important considerations in studying polymer degradation.

The *physical state* of the material is decisive with respect to the possible alterations caused by mechanical forces. In solid polymeric bodies, *morphology, orientation, sample geometry,* and *initial particle size* strongly influence the degradation process. The presence of rubbery particles (in suitable amount and

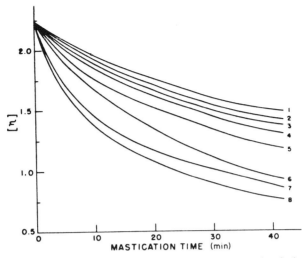

Figure 9.9. The effect of various solvents on the decrease of viscosity during polystyrene mastication (at 40°C, 30 vol % solvent, plough scrolls at 120 rpm, under nitrogen): (1) toluene, (2) ethylene dichloride, (3) methyl ethyl ketone, (4) acetone, (5) benzene, (6) n-butyl acetate, (7) carbon tetrachloride, (8) petroleum ether. (Reprinted, by permission, from reference 9.6.)

distribution) in a glassy polymer matrix may have a toughening effect on the brittle polymer. The "high impact modifiers" function by acting as energy absorbers or crack stoppers. The shear stability of polymer solutions depends on the *concentration*, or *polymer-solvent interaction*. The effect of various solvents on polystyrene mastication is illustrated in Figure 9.9. Gases or liquids — even when present only in small amounts — may influence the degradation mechanism; such *environmental* factors are very important in practical applications.

Time and temperature, of course, strongly affect the extent of mechanical degradation. While increasing treatment time always increases degradation, pure mechanical degradation processes usually have a negative temperature coefficient, as illustrated in Figure 9.10 for polyisobutylene milling. Although this behavior can easily be explained by the increase of rigidity (in solids) or of viscosity (in melts) with decreasing temperatures, stress-induced scission processes having positive temperature coefficients are also observed. A possible explanation is that at higher temperatures a broader distribution of chain length is available for scission.

Pressure induces bond rupture only when applied at high values such as 10^3–10^5 atm; however, relatively low pressures influence ultrasonic degradation. As shown in Figure 9.11, the rate constant of viscosity decrease during ultrasonic irradiation of a polystyrene/benzene solution passes through a maximum with

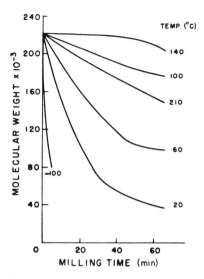

Figure 9.10. The effect of temperature on the decrease of molecular weight during polyisobutylene milling. (Reprinted, by permission, from reference 9.6.)

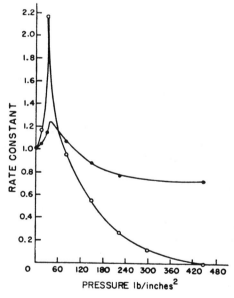

Figure 9.11. Rate constant of degradation (viscosity decrease) during ultrasonic irradiation (500 kHz, 10 W/cm^2) of polystyrene in benzene as a function of excess pressure, exerted by a gas above the solutions (○), or via a mercury column direct on the solutions (●). (Reprinted, by permission, from reference 9.6.)

increasing pressure. Pressure increase suppresses cavitation which plays an important role in ultrasonic degradation.

It is obvious that *stress* and *strain* are among the most important parameters of a mechanical degradation experiment. However, the stress-strain behavior of various materials in various physical states, the effect of the strain rate, loading rate, and shearing rate, the role of stress distribution in the sample, the influence of stress-strain history on the deformation or fracture process, and the specific effects of fatigue are all subjects of extensive studies on the mechanical properties, fracture mechanics, rheology, and thermodynamics of polymers, and cannot be handled within the scope of degradation treatment. On the other hand, complete understanding of mechanical degradation and effective utilization of its possibilities require fundamental knowledge of these domains.

REFERENCES

9.1. Andrews, E. H. and Reed, P. E. Molecular fracture in polymers. *Advances in Polymer Science* 27:1 (1978).

9.2. Baramboim, N. K. *Mechanochemistry of Polymers*. London: Maclaren, 1964.

9.3. Becht, J. and Fischer, H. Elektronenspinresonanz-, Differentialthermoanalyse- und Molekulargewichtsuntersuchungen an mechanisch beanspruchten Kunststoff-Fäden. *Kolloid-Z. u. Z. Polymere* 240:766 (1970).

9.4. Brett, H. W. W. and Jellinek, H. H. G. Degradation of long-chain molecules by ultrasonic waves. VI. Effect of pressure. *J. Polymer Sci.* 21:533 (1956).

9.5. Bueche, F. *Physical Properties of Polymers*. New York: Interscience, 1962.

9.6. Casale, A. and Porter, R. S. *Polymer Stress Reactions*, Vols. 1 and 2. New York: Academic Press, 1979.

9.7. Ceresa, R. J. The mechano-chemical degradation of plastomers. *Plastics Inst. Trans. J.* 28:2 (1960).

9.8. Dobrodumov, A. V. and Elyashevich, A. M. Simulation of brittle fracture of polymers by a network model in the Monte-Carlo method. *Soviet Physics Solid State* 15:1259 (1973).

9.9. Johnsen, U. and Klinkenberg, D. ESR-Messung von Kettenbrucken in mechanisch beanspruchtem Polyamid. *Kolloid-Z. u. Z. Polymere* 251:843 (1973).

9.10. Kausch, H. H. *Polymer Fracture*. Heidelberg: Springer, 1978.

9.11. Pazonyi, T., Tüdős, F. and Dimitrov, M. Untersuchung der beim Schnitzeln von Polymeren sich bildenden Radikale. *Angew. Makromol. Chem.* 10:75 (1970).

9.12. Peterlin, A. Chain scission and plastic deformation in the strained crystalline polymer. *J. Polymer Sci.* C32:297 (1971).

9.13. Porter, R. S., Cantow, M. J. R. and Johnson, J. F. Polymer degradation. VI. Distribution changes on polyisobutene degradation in laminar flow. *Polymer* 8:87 (1967).

9.14. Simionescu, C. L., Vasiliu-Oprea, C. and Negulianu, C. Mechanochemical syntheses. *J. Polymer Sci.: Polymer Symposia* 64:149 (1978).

9.15. Sohma, J. Electron spin resonance studies of the mechanical degradation of polymers. In (Grassie, N. ed.) *Developments in Polymer Degradation–2*. London: Applied Science Publishers, 1979.

9.16. Sohma, J. and Sakaguchi, M. ESR studies on polymer radicals produced by mechanical destruction and their reactivity. *Advances in Polymer Science* **20**:109 (1976).

9.17. Watson, W. F. Mechanochemical reactions. In (Fettes, E. M., ed.) *Chemical Reactions of Polymers*. New York: Wiley-Interscience, 1964.

9.18. Zhurkov, S. N., Zakrevskii, V. A., Korsukov, V. E. and Kuksenko, V. S. Mechanism leading to the development of submicroscopic fissures in stressed polymers. *Soviet Physics Solid State* **13**:1680 (1972).

Chapter 10
Stabilization of Polymers

As has been shown in the preceding chapters, most polymers are more or less susceptible to various forms of degradation. This inclination may lead to substantial damage of the material and deterioration of its useful properties, and is an important problem in polymers produced on a large scale. Thus, stabilization of polymers is of great economic importance, and techniques to accomplish this are as old as the production of polymers themselves. The mass production of PVC would never have developed to its present degree without the development of effective stabilizers for this normally very unstable polymer.

The aim of stabilization is to avert and/or neutralize the factors, and to prevent and/or control the processes, which may damage a polymer during its preparation, compounding, processing, or use. With respect to the various applications of polymers, the purpose of stabilization is to maintain the original characteristics of the product and assure its desired service life.

A distinction is customarily made between preventive and arrestive stabilization. A *preventive* measure results in the production of a more stable polymer. In addition to the manufacture of special polymers having peculiar composition and structure (e.g., ring systems in the chain which cause backbone rigidity, silicones, coordination polymers, etc.), preventive measures can also be taken during the production of conventional polymers. In order to produce a more stable material (internal stabilization), monomers of higher purity may be used. To avoid initiator residue in the polymer, radiation polymerization can be applied. Cleaner material can also be obtained by better isolation of the polymer after termination. Another method of internal stabilization is the copolymerization with a few percent of comonomers (e.g., with dioxolane in the case of polyoxymethylene to decrease depolymerization).

Arrestive stabilization can be carried out by the removal, neutralization, or inactivation of various potential degradation sources, defects, etc., which accumulate in a polymer. The introduction of reactive structures into the polymer chain, for example, side groups which act as antioxidants, may also arrest degradation processes (built-in stabilizers).

The primary method of arrestive stabilization, however, is the introduction of reactive species into the polymer (external stabilization). These additives (stabilizers) may act in various ways depending on their chemical nature and on the application conditions. In the following discussion we will consider only the latter type of stabilization.

10.1 GENERAL REQUIREMENTS

Besides having stabilizing characteristics, which will be discussed later for specific classes, stabilizers must satisfy some very important general requirements. When selecting a stabilizer or a combination of stabilizers, we must systematically eliminate alternatives having unwanted characteristics and experimentally test the choice in each approach. Sometimes, we must sacrifice an otherwise excellently performing stabilizer, e.g., when it does not correspond to some public health regulations (in the United States, the FDA, the Food and Drug Administration, sanctions additives for food contact applications). In many cases, a compromise has to be found between performance criteria and restraints of applicability.

The stabilizer itself must be *stable*; its thermal and light stability, and its stability against moisture, oxygen, and living organisms, are important considerations. Which of these requirements is dominant, depends on the application conditions. The stabilizer must usually be *nonvolatile*, *tasteless*, and *odorless*. Neither the additive nor its transformation products may *stain* the polymer matrix (however, many applications tolerate colored or coloring stabilizers). As previously mentioned, in food contact applications the stabilizer must meet *FDA requirements*; however, *toxicity* is also important with respect to the safety of workers processing the material.

A good stabilizer is *nonmigrating*. Migration to the surface of the material would cause (e.g., in the case of soft PVC car seat covers) undesirable *exudation* (spewing). Similarly, an unwanted property of some PVC stabilizers is *plating* (plateout) which results in sticky deposits on the metal surfaces of processing equipment. The *extractability* of the stabilizer by water and a propensity for formation of *water fogs* or for development of *haze* (opacity) are disadvantageous stabilizer properties.

Compatibility with the polymers and with other additives, and the lack of *chemical interactions* with them, are important requirements. Before selecting a stabilizer, it is also essential to know its *effects on the physical characteristics of the polymer*, e.g., on rheological behavior (plasticizing and lubricating action), on mechanical and electrical properties, etc.

Last, but not least, the *cost* of the stabilizer should be mentioned. Some excellent stabilizers are very expensive materials; others are inexpensive but not so effective. In comparisons, the costs of stabilizers which have comparable activity must be considered. Despite the relatively small amount of stabilizer in a polymer compound, its cost contribution may be quite high. The search for synergistic stabilizer combinations (i.e., for combinations performing more effectively than would be expected on the basis of additivity) is sometimes motivated by financial considerations.

10.2 ANTIOXIDANTS

Because of its general occurrence in every phase of a material's life, the oxidation of polymers is probably the most important of the various deteriorating processes. Therefore, the stabilizers used to retard or inhibit this chain reaction (for a detailed description of which see Chapter 6), i.e., antioxidants, find wide application in many stabilizer packages. Two main groups of antioxidants are distinguished according to their mode of action: primary or chain breaking, and secondary or preventive antioxidants.

Chain Breaking Antioxidants

Chain breaking antioxidants interfere with the chain propagation steps [reactions (III.1) and (III.2) in Figures 6.1, 6.4, and 6.5] of polymer oxidation. These antioxidants can terminate the kinetic chain when they are *radical species* (stable free radicals, e.g., diaryl nitroxides):

$$RO_2 \cdot + In \cdot \longrightarrow \text{nonradical products} \qquad (10.1)$$

where $In \cdot$ stands for the stable free radical inhibitor. Phenoxy radicals (obtained, for example, from tri-*t*-butylphenol by oxidation with PbO_2), as well as stable radicals of naphthols and aromatic amines (the latter are precursors of nitroxide radicals), are examples of stable free radical scavengers. Diphenylpicrylhydrazyl, already mentioned (see section 9.1) as a scavenger of mechano-radicals, is frequently employed.

Quinones and *conjugated molecules* (e.g., anthracene) can also function as free radical traps (scavengers), converting the $RO_2 \cdot$ radical to a more stable free radical which is less capable of continuing the oxidation chain reaction. Quinones can be formed as transformation products of a phenolic antioxidant [see reactions (10.9) through (10.12)] and can extend the active life span of the latter. As a polyconjugated system, carbon black can inhibit oxidation and also functions as a light stabilizer; however, its combination with phenolic and amine compounds is disadvantageous because it catalyzes their oxidation.

The most commonly used chain terminating antioxidants are *hydrogen donors*. Inhibitors of the InH type interfere with oxidative chain propagation by competing with the polymer for the peroxy radicals:

$$RO_2 \cdot + InH \longrightarrow RO_2H + In \cdot \qquad (10.2)$$

where In· is a more stable radical. Replacement of the very reactive RO_2· by the unreactive In· retards the chain reaction and slows the oxidation. Under favorable conditions, these stable radicals will react only with the active ones, causing their further withdrawal according to the termination reaction (10.1), and will not participate in undesirable reactions. The extent of side reactions influences the efficiency of inhibition. The stoichiometric coefficient (μ) of the antioxidant is defined as the number of RO_2· radicals terminated by one molecule of inhibitor. This coefficient depends on the inhibited system and is 2 in the case of a single reaction site (e.g., one OH group) when no side reaction takes place. Such unwanted reactions are, for example, transfer to the polymer:

$$In \cdot + RH \longrightarrow InH + R \cdot \qquad (10.3)$$

or to the hydroperoxides:

$$In \cdot + RO_2H \longrightarrow InH + RO_2 \cdot \qquad (10.4)$$

These reactions are undesirable because they would regenerate active radicals. Fortunately, these reactions can be considerably suppressed by using sterically hindered aromatic compounds containing relatively large substituents (t-butyl groups) in the ortho position to the hydrogen-donating group.

The length of the induction period of oxidation caused by a chain breaking antioxidant (inhibition period) can be calculated as follows. When W_{11} is the rate of the initiation reaction [(I.1) in Figure 6.5], X the concentration of the RO_2· radical, and k_{cb} the rate constant of the chain breaking reaction (10.2), the following differential equations hold:

$$\frac{dX}{dt} = 2W_{11} - \mu k_{cb} X[InH] \qquad (10.5)$$

$$\frac{d[InH]}{dt} = -k_{cb} X[InH] \qquad (10.6)$$

where μ is the stoichiometric coefficient of the antioxidant in the inhibition reaction under investigation. Substituting Eq. (10.6) into Eq. (10.5) and integrating lead to:

$$X = 2W_{11}t - \mu([InH]_0 - [InH]) \qquad (10.7)$$

Extrapolation of $X(t)$ after the consumption of the antioxidant ($[InH] = 0$) gives the length of the inhibition period:

$$t_{inh} = \frac{\mu}{2W_{11}}[InH]_0 \qquad (10.8)$$

i.e., t_{inh} is proportional to the initial concentration of the chain breaking antioxidant. The most effective members of this antioxidant group are hindered phenols and secondary aryl amines, containing reactive O—H and N—H bonds, respectively.

Phenolic antioxidants are commercially available in a wide range of molecular weights. Butylated hydroxytoluene (BHT, see Table 10.1) is a low molecular weight nonstaining phenolic compound which is probably the most widely used antioxidant. The transformations of BHT during application in an oxidizing system have been extensively studied. The In· radical formed in the reaction

$$+ RO_2{}^{\bullet} \rightarrow \qquad + RO_2H \qquad (10.9)$$

(InH = BHT)   (In·)

can transform to a mesomeric quinoid-type radical (In'·)

$$\longleftrightarrow \qquad (10.10)$$

(In·)   (In'·)

which consumes another $RO_2 \cdot$ radical:

Table 10.1. Registered Trade Name and Chemical Composition of Some Important Chain Breaking Antioxidants.

CHEMICAL COMPOSITION	REGISTERED TRADE NAME	SUPPLIER
	BHT Ionol Topanol OC	Uniroyal Shell ICI
$CH_2-CH_2-COO-C_{18}H_{37}$	Irganox 1076	Ciba-Geigy
	Cyanox 2246	American Cyanamid
	Irganox 1010	Ciba-Geigy

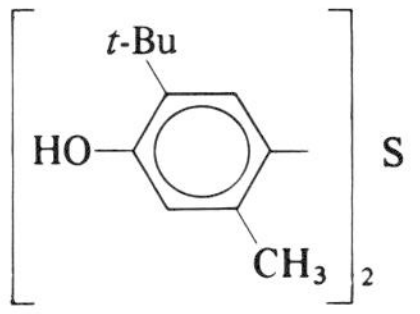

$$\left[\begin{array}{c} \text{HO} \quad t\text{-Bu} \\ \text{CH}_3 \end{array} \right]_2 S$$

Santonox R   Monsanto

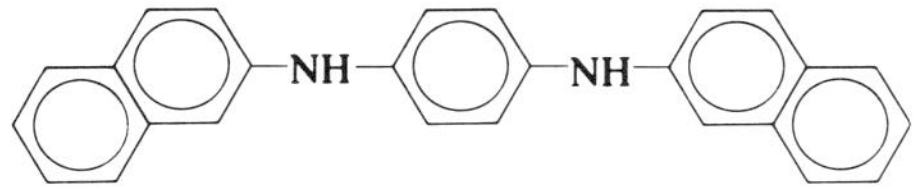

Nonox CI   ICI

$$\text{(In}'\cdot\text{)} + RO_2^{\boldsymbol{\cdot}} \longrightarrow \text{(In}'-OOR)} \tag{10.11}$$

There are, however, many possibilities of further reactions (e.g., isomerization of In· to a benzyl radical) which lead to a very complex assembly of products. The In· radical is also able to disproportionate, regenerating InH and forming a quinone methide (QM):

$$\text{(In}\cdot\text{)} \longrightarrow \text{(InH)} + \text{(QM)} \tag{10.12}$$

Formation of products with highly colored quinomethinoid and quinonoid structures is a typical disadvantage of some phenolic antioxidants.

Although BHT is probably one of the most effective hindered phenol antioxidants, it has the drawback of being highly volatile. Higher molecular weight monophenols are better in this respect; bis-, tris-, and polyphenols have, besides a higher molecular weight, the advantage of an increased number of reactive groups. A few examples (Irganox 1010 and 1076, Cyanox 2246) are shown in Table 10.1.

Some thiophenol compounds made very effective, clear, nonstaining, light stable antioxidants (e.g., Santonox R). Examples are listed in Table 10.1.

The antioxidant action of secondary aryl amines, an example of which is given in Table 10.1 (Nonox CI), can be explained by the same mechanism as that of the phenols. However, an interesting additional possibility is the formation of nitroxy radicals which may function as radical scavengers. For example:

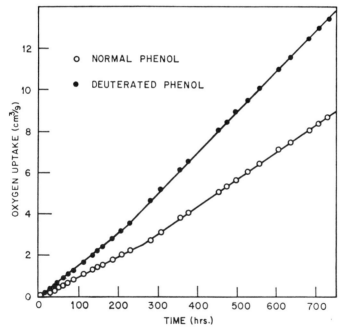

$$+ \ 2RO_2\cdot$$

$$\longrightarrow \quad + \ 2RO_2H$$

(10.13)

$$\longrightarrow \quad + \ 2ROH$$

Some of the amine antioxidants not only are able to act as primary antioxidants but also are capable of peroxide decomposition, especially at higher temperatures. A disadvantage is that amines usually cause strong discoloration. Therefore, amine antioxidants are used mainly in applications where high temperature production is important but color is not, e.g., in rubber products, nylon tire cords, etc. Amines which cause only slight coloration are increasingly applied in

Figure 10.1. Oxidation of purified polyisoprene in the presence of 2 pph normal and deuterated BHT (butylated hydroxytoluene) antioxidant, respectively (75°C, 1 atm O_2, 8.82 × 10^{-5} mol/g additive). (Reprinted with permission from reference 10.20. Copyright 1963, American Chemical Society.)

polyesters and polyethers. Some hindered amines are not only antioxidants but also excellent UV stabilizers (hindered amine light stabilizers = HALS). These compounds (various piperidine derivatives) will be dealt with among the photo-stabilizers.

The mechanism of inhibition of hindered phenols and aryl amines (i.e., reaction via active hydrogen) has been demonstrated by showing the isotope effect caused by substituting deuterium for the active hydrogens of BHT (Figure 10.1) and N-phenyl-2-naphthylamine (Figure 10.2). As can be seen, a significant decrease in effectiveness was obtained in both cases.

Preventive Antioxidants

Preventive antioxidants can prevent oxidative deterioration by counteracting the formation of free radicals. Various light absorbers which function as UV stabilizers, screening out the dangerous radiation or quenching the excited species, act as preventive antioxidants as well. Another group of these stabilizers exerting antioxidant effects function as *metal-ion deactivators*. The capacity of some metal ions for oxidation-reduction and complex formation is well known; they can exert a pro-oxidant effect, e.g., by promoting the radical decomposition of

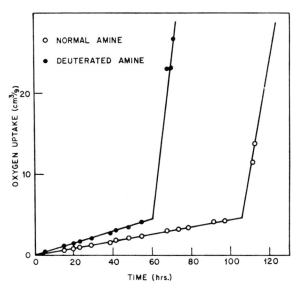

Figure 10.2. Oxidation of purified polyisoprene in the presence of 1 pph normal and deuterated N-phenyl-2-naphthylamine, respectively (75°C, 1 atm O_2, 4.41 × 10^{-5} mol/g additive). (Reprinted with permission from reference 10.20. Copyright 1963, American Chemical Society.)

hydroperoxides:

$$RO_2H + M^{(n)+} \rightarrow RO_2\cdot + M^{(n+1)+} + HO^- \qquad (10.14)$$

$$RO_2H + M^{(n+1)+} \rightarrow RO_2\cdot + M^{(n)+} + H^+ \qquad (10.15)$$

Some compounds, capable of metal chelate formation, may neutralize this free radical sources.   For example, a bishydrazone can deactivate copper by chelation:

$$(10.16)$$

In the chelate formed, Cu is coupled by covalent bonds to the oxygen atoms and by coordinate links to the nitrogens.   Chelate forming compounds containing reactive OH or NH groups can also function as chain breaking antioxidants, e.g.,

$$(10.17)$$

Various compounds containing phosphorus or sulfur are able to destroy hydroperoxides without the formation of free radicals. *Hydroperoxide decomposing* antioxidants are very important because hydroperoxides are the main radical sources in polymer autoxidation.   Some hydroperoxide decomposers such as zinc dibutyldithiocarbamate and hindered piperidine derivatives (HALS) can function as light stabilizers as well.   Among the phosphorus compounds, organic phosphites are effective nondiscoloring antioxidants which function by reducing the hydroperoxides to alcohols.   The phosphite itself transforms to phosphate. A triaryl (e.g., trisnonylphenyl) phosphite functions as follows:

$$RO_2H + (ArO)_3P \rightarrow ROH + (ArO)_3PO \qquad (10.18)$$

Some of the phosphite antioxidants are sensitive to hydrolysis.   When used in combination with other stabilizers (e.g., with phenolic antioxidants or with solid Ba/Cd, liquid Ba/Cd, and Ca/Zn PVC heat stabilizers), they can affect the moisture resistance of the material. The use of phosphites in stabilizer packages is often advantageous because they can inhibit the formation of discoloring quinonoid structures from hindered phenols.

Thioesters are also frequently used together with phenolic antioxidants. Their combination usually exerts a synergistic stabilization effect. An example of the interaction of thioethers with hydroperoxides was shown in reaction (6.2). In polyolefin stabilization, one of the most widely used thioethers is dilauryl thiodipropionate:

$$\left[H_{25}C_{12}O - \overset{\overset{\displaystyle O}{\|}}{C} - CH_2 CH_2 - \right]_2 S \tag{10.19}$$

The length of the induction period caused by hydroperoxide decomposer can be obtained from the differential equations

$$\frac{dX}{dt} = 2W_{11} \tag{10.20}$$

$$\frac{dY}{dt} = k_{31}X - k_{hd}Y[HD] \tag{10.21}$$

$$\frac{d[HD]}{dt} = -k_{hd}Y[HD] \tag{10.22}$$

where W_{11} is the rate of the initiation reaction [(I.1) in Figure 6.5] X, Y, and [HD] represent the concentrations of $RO_2\cdot$, RO_2H, and the hydroperoxide decomposer, respectively; and k_{31} and k_{hd} denote the rate constants of reaction (III.1) in Figure 6.5 and of the hydroperoxide decomposing reaction [e.g., reaction (10.18)], respectively. Integration of Eq. (10.20) gives

$$X = 2W_{11}t \tag{10.23}$$

The substitution of Eqs. (10.22) and (10.23) into Eq. (10.21), followed by integration, leads to

$$Y = k_{31}W_{11}t^2 - ([HD]_0 - [HD]) \tag{10.24}$$

Extrapolation of $Y(t)$ after the consumption of the antioxidant ([HD] = 0) gives the length of the inhibition period:

$$t_{inh} = \frac{1}{\sqrt{k_{31}W_{11}}} ([HD]_0)^{1/2} \qquad (10.25)$$

i.e., we obtained a one-half order dependence on the initial antioxidant concentration, contrary to the first order obtained for chain breaking antioxidants.

The above considerations concerning the length of induction periods in both types of antioxidants are applicable only when the additives are very active, i.e., when the participation of the hydroperoxide in reaction (IV) of Figure 6.5 (degenerate chain branching) is negligible compared to the nonradical decomposition reaction. The amount of additive must be above a critical concentration; below this concentration, as shown by Shlyapnikov and others, the antioxidant cannot prevent the acceleration of oxidation.

It should be mentioned that some antioxidants not only terminate chains or decompose hydroperoxides but may take part in the initiation of oxidation as well. For example, a direct oxidation of InH with molecular oxygen

$$O_2 + InH \longrightarrow HO_2\cdot + In\cdot \qquad (10.26)$$

is usually undesirable. Because of its higher activation energy, the role of reaction (10.26) compared to that of the chain breaking reaction (10.2) increases with increasing temperature. For example, in melt stabilization of polypropylene with BHT, the majority of the inhibitor is consumed by direct oxidation of the phenolic antioxidant. This can, however, be advantageous under processing conditions where only partial replacement of the consumed oxygen is possible; BHT can effectively remove oxygen from the melt, thus also preventing polymer oxidation.

10.3 PHOTOSTABILIZATION

Stabilization against photodegradation of polymers is very important, particularly in outdoor applications. As shown in Chapter 7, the actinic UV component of sunlight, acting for long times on polymeric bodies and usually accompanied by other damaging factors such as moisture and oxygen, causes irreversible chemical and physical changes in the material, leading to deterioration of its properties.

The possibilities of stabilization by photostabilizers are illustrated in Table 10.2. As can be seen, there are several modes of stabilizing action corresponding to various phases of the photodegradation process. Table 10.2 shows a photooxidative process in which, in addition to the photostabilizers, both types of antioxidants can be beneficial. This is an example which illustrates that in practice, combinations of stabilizers (stabilizer packages) are required for adequate protection.

Table 10.2. Possibilities of Photostabilization.

PHOTOOXIDATIVE PROCESS	TYPE OF STABILIZER
UV light	← ─ ─ ─ ─ screens (surface action) ← ─ ─ ─ ─ UV absorbers
↓	
polymer	
↓	
excited polymer	← ─ ─ ─ ─ quenchers
singlet oxygen	← ─ ─ ─ ─ singlet oxygen quenchers
free macro-radicals	← ─ ─ ─ free radical scavengers
hydroperoxides	← ─ ─ ─ peroxide decomposers
↓	
deteriorated polymer	

Light Screens

Reflecting or absorbing incident light before it reaches the polymer surface is the trivial solution to photostabilization. Protective *coatings* can function this way. A layer of aluminum powder on a polyethylene or acrylonitrile-butadiene-styrene (ABS) resin surface reflects incident light. This kind of protection has not been widely used for synthetic polymers. For natural polymers, the use of coatings is a conventional method of treatment; for example, the painting of wood for protection against light (and other damaging effects) is widely applied.

Pigments, dispersed in the polymer matrix function by limiting the penetration of UV radiation into the interior of the material. Both inorganic and organic compounds are widely used in the plastics industry as pigments. A traditional pigment is carbon black which can act not only as an inner screen but also as a free radical scavenger and quencher. Various types of carbon black with different particle size distributions and different physical properties, depending on their preparation conditions (channel black, furnace black, lampblack), are used. Their effectiveness as stabilizers, like that of other pigments, strongly depends on their adequate dispersion in the polymer matrix.

Among the inorganic pigments, zinc oxide is widely used in various applications as an effective photostabilizer. The white pigment has excellent reflection at 240-380 nm. Colored organic pigments exhibit good ultraviolet absorption properties. Some dyes, e.g., phthalocyanine blues and greens, are advantageously applied pigments.

Ultraviolet Absorbers

Ultraviolet absorbers include those additives which absorb and dissipate the energy of UV radiation which has penetrated into the bulk polymer. Strictly speaking, some pigments classified as light screens should be treated here. The distinction is partly conventional, partly based on the fact that pigments are not transparent in the visible region.

Although many organic compounds absorb UV light, only those which are able to dissipate the absorbed energy without their own deterioration, e.g., by radiationless internal conversion or by intersystem crossing, etc. (see section 7.1), can be used as photostabilizers. UV absorbers in which intramolecular hydrogen bonding occurs, are assumed to undergo deactivation of their excited state by internal conversion, i.e., without changing spin multiplicity.

2-Hydroxybenzophenone and its derivatives (e.g., Cyasorb UV 531; see Table 10.3) can be effectively applied in almost all kinds of polymers. These compounds function not only by UV absorption but also by quenching excited states. Because of the *o*-hydroxy–ketone configuration, metal chelation is possible. A typical performance of this photostabilizer is illustrated in Figure 10.3 which shows the effect of 1 wt % Cyasorb UV 531 on the formation of carbonyl and hydroperoxide groups in polypropylene during UV irradiation. The decrease in the amount of stabilizer (as measured by UV spectrophotometry) is also shown in the figure. The UV absorber concentration decreases linearly during irradiation. At the end of the induction period, there is still about 25% of the added stabilizer present.

It should be noted that while the ortho isomers are excellent photostabilizers, *p*-hydroxybenzophenones sensitize photodegradation. A limitation of the use of *o*-hydroxybenzophenones is that they may cause discoloration.

Salicylaldehyde and salicylates (e.g., phenyl salicylate) behave like the benzophenones.

2-Hydroxyphenylbenzotriazole and derivatives (e.g., Tinuvin 326; see Table 10.3), which have the possibility of internal hydrogen bonding, are excellent UV absorbers. Benzotriazoles are not as effective as benzophenones in polyolefins but are more effective in polycarbonates. The performance of Tinuvin in polypropylene is illustrated in Figure 10.4. As can be seen, the effectiveness of Tinuvin 326 is far from that of Cyasorb UV 531. The decrease in stabilizer concentration and the formation of carbonyl groups begin at about the same time.

Table 10.3. Registered Trade Name and Chemical Composition of Some Important Photostabilizers.

CHEMICAL COMPOSITION	REGISTERED TRADE NAME	SUPPLIER
	Cyasorb UV 531 Mark 1413 AM-300	American Cyanamid Argus Ferro
	Tinuvin 326	Ciba-Geigy
	NBC AM-104	DuPont Ferro
	Tinuvin 770	Ciba-Geigy

Many other types of organic compounds can function as UV absorbers (e.g., the phenyl esters of benzoic acid, some substituted acrylonitrile derivatives, ferrocene derivatives); however, their practical importance is less than that of the benzophenones and benzotriazoles.

Quenchers

Quenching is energy transfer between an excited molecule and a photostabilizer. As shown in Table 10.2, this type of deactivation is possible with respect to both

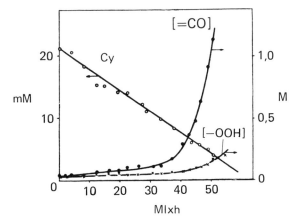

Figure 10.3. Change of the concentration of carbonyl ([$>$CO]) and hydroperoxide ([−OOH]) groups, as well as of the added (1 wt %) stabilizer (Cyasorb UV 531; Cy) in polypropylene during UV irradiation by a xenon lamp (dose in megaluxhours, Mlxh).

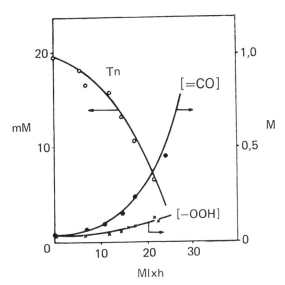

Figure 10.4. Carbonyl ([$>$CO]) and hydroperoxide ([−OOH]) formation, and decrease of the amount of added (1 wt %) stabilizer (Tinuvin 326; Tn) in polypropylene during xenon lamp irradiation (dose in megaluxhours, Mlxh).

the excited polymer molecule and the singlet oxygen. Quenching may occur either via an excited quencher molecule

$$A^* + Q \longrightarrow A + Q^* \longrightarrow\longrightarrow A + Q \tag{10.27}$$

or via the formation of an excited complex

$$A^* + Q \longrightarrow [A \ldots Q]^* \longrightarrow\longrightarrow A + Q \tag{10.28}$$

In both cases, the quencher will be deactivated (regenerated) by some photophysical process(es).

Singlet oxygen can be effectively quenched by olefins, aliphatic amines (e.g., triethylamine), and 2,5-dimethylfuran. These compounds, however, scarcely find practical application in polymer photostabilization. Metal chelates, particularly some *nickel-organic compounds* are far more important quenchers, being effective in both singlet oxygen and excited polymer quenching. Nickel dibutyldithiocarbamate (NBC; see Table 10.3) is a very versatile photostabilizer because it functions as a UV absorber and as a hydroperoxide decomposer. The corresponding zinc compound which does not cause discoloration, in contrast to NBC which imparts a greenish tint to the polymer, is a less effective quencher but a good peroxide decomposer. The photostabilization efficiency of NBC is illustrated in Figure 10.5. Carbonyl and hydroperoxide formation, together with the decrease of the amount of stabilizer, are shown as a function of the irradiation dose. The induction period is about twice as long as with Cyasorb UV 531 (Figure 10.3), and oxidation begins when NBC is consumed.

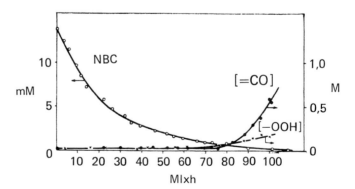

Figure 10.5. Carbonyl ([>CO]) and hydroperoxide ([−OOH]) formation and decrease of the amount of the added (1 wt %) stabilizer (nickel dibutyldithiocarbamate, NBC) in polypropylene as a function of the dose (megaluxhours, Mlxh) of xenon lamp irradiation.

Very effective, relatively new photostabilizers are the *hindered amine light stabilizers* (HALS). Commercially available HALS compounds are piperidine derivaties [e.g., 4-(2,2,6,6-tetramethylpiperidinyl) sebacate: Tinuvin 770, see Table 10.3]. They absorb UV light only below 270 nm, thus their photostabilizing activity cannot be interpreted by UV absorption. They exert an excellent stabilizing effect, 2–4 times greater than the best stabilizer previously known. This is illustrated in Figure 10.6. As can be seen, the carbonyl and hydroperoxide concentration of the polymer remained at the original level until the total mechanical breakdown of the samples at an irradiation dosage of approximately 210 Mlxh (megaluxhours). The amount of additive decreased according to an exponential curve. The stabilizer was practically consumed (or transformed) when the samples broke into fragments.

The major function of HALS compounds is radical scavenging. Thus, they are chain breaking antioxidants. Their ability to protect polymers against UV irradiation distinguishes them from conventional antioxidants: they must exert other types of protective effects as well.

The radical scavenging effect is due to stable nitroxyl radicals which are formed in the sterically hindered piperidine derivatives during their spontaneous

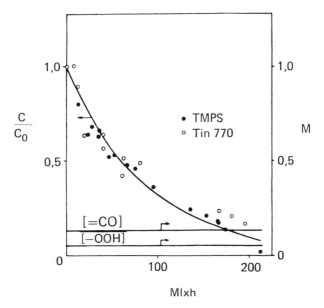

Figure 10.6. Carbonyl ([>CO]) and hydroperoxide ([−OOH]) content of polypropylene, and decrease of the relative amount (C/C_0) of stabilizer as a function of xenon lamp irradiation dose (megaluxhours, Mlxh). HALS compounds, added in 1 wt %, were Tinuvin 770 (Ciba-Geigy) and TMPS (the same compound, a laboratory product).

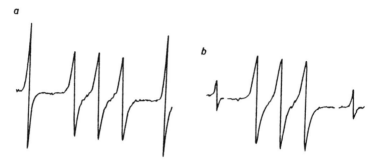

Figure 10.7. ESR spectra of 4-(2,2,6,6-tetramethylpiperidinyl) sebacate (TMPS; a) and of the 2,2,6,6-tetramethylpiperidine-N-oxy radical (b), both laboratory products, dissolved in toluene, at room temperature. Concentration: (a) ~0.1 g/ml, (b) ~10^{-4} g/ml. Amplification: (a) 1400×, (b) 450×. The extreme lines belong to Mn used as reference. (Reprinted, by permission of Applied Science Publishers Ltd., from reference 10.7.)

oxidation in air. This can be seen from a comparison of the ESR spectra of the piperidine derivative and of the corresponding piperidine-N-oxy radical (Figure 10.7). During the irradiation of the polymer the concentration of nitroxyl radicals first increases then, after reaching a maximum, decreases (Figure 10.8). The kinetics of N-oxy radical concentration change, typical of reaction intermediates, may be attributed to the formation of radicals from HALS and to their reaction with $RO_2 \cdot$ radicals of the polymer oxidation.

It should be noted that HALS compounds (like any amine stabilizers) cannot be used in PVC because they would enhance dehydrochlorination.

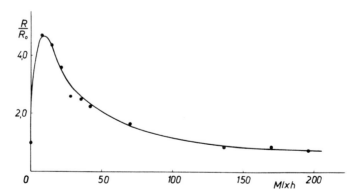

Figure 10.8. Change of the relative concentration (R/R_0) of nitroxyl radicals as a function of xenon lamp irradiation dose (megaluxhours, Mlxh) in polypropylene containing 1 wt % TMPS. (Reprinted, by permission of Applied Science Publishers Ltd., from reference 10.7.)

10.4   HEAT STABILIZATION OF PVC

Polyvinyl chloride, one of the most important commercial polymers, is very unstable. Thus, efforts to improve its stability, and hence extend its useful life-time, are as old as the polymer itself. Most of the early work done in this field was completely empirical; nevertheless, many of the stabilizers, suggested without understanding their mode of action, are very effective and are still in use. PVC stabilization has an unusually abundant patent literature, and hundreds of stabilizers, sometimes differing only slightly from each other, are now commercially available.

Internal stabilization of PVC using selected comonomers which may interrupt the unzipping chain dehydrochlorination or act as built-in stabilizers has not found wide application. Another preventive technique suggested by many authors is pretreatment of the polymer to increase its intrinsic stability by removing structural defects or by substituting a stable group for a labile one. Although defective sites containing labile chlorine atoms, e.g., allylic chlorines, are in fact important sources of PVC degradation, pretreatment techniques are presently too expensive for practical purposes because such techniques usually require the dissolution of the polymer. Such procedures, which result in polymers with enhanced stability, include treatment of the polymer with alkyl-aluminum compounds (in the presence or absence of monomers), lithium aluminum hydride, diphenylmethyllithium, or dialkyltin mercaptides.

The conventional technique for PVC stabilization is the addition of a stabilizer or a combination of stabilizers to the polymer. Most PVC heat stabilizers are organic metal salts having the composition MY_2 where M is a bivalent cation of metals such as Ca, Zn, Ba, Cd, Pb, or a dialkyltin cation, etc., and Y is an organic anion such as carboxylate, alcoholate, or thiolate. Most traditional "stabilizers" function as *hydrogen chloride acceptors*:

$$HCl + MY_2 \longrightarrow MYCl + HY \qquad (10.29)$$

$$HCl + MYCl \longrightarrow MCl_2 + HY \qquad (10.30)$$

These reactions seem at first to represent only an illusion of stabilization for while an induction period occurs with respect to HCl generation, the polymer still discolors. However, because HCl in many cases does promote the degradation of PVC, its removal may be regarded as true stabilization. Thus, the above reactions are indeed stabilization reactions.

The most effective stabilizers, as first recognized by Frye and Horst (1959), are those which are able to replace labile chlorine atoms by stable groups:

$$RCl + MY_2 \longrightarrow MYCl + RY \qquad (10.31)$$

$$RCl + MYCl \longrightarrow MCl_2 + RY \qquad (10.32)$$

This mode of action is a true stabilization because the conversion of defects into stable structures prevents dehydrochlorination and polyene formation which would start at these defect sites. It should be noted that stabilizers which function according to the *Frye-Horst mechanism* (e.g., organotin compounds) may act as HCl acceptors as well. Thus, they can be partly consumed by reaction (10.29) or (10.30) before being able to exert their more effective action.

In some cases, MCl_2 or $MYCl$ type metal chlorides formed in the above reactions catalyze PVC dehydrochlorination. The extent of this undesired behavior is determined by the chemical character of the metal ion in question. The use of *mixed stabilizers* such as Ba/Cd or Ca/Zn stearates can minimize the catalytic effect of MCl_2; this can be explained by an exchange reaction between the two metals:

$$M^{(1)}Cl_2 + M^{(2)}Y_2 \longrightarrow M^{(1)}YCl + M^{(2)}YCl \qquad (10.33)$$

For example cadmium stearate is more active in the labile chlorine exchange reaction than barium stearate (see Figure 10.11). Thus, unreacted barium salt is still present when the cadmium salt has been transformed to $CdCl_2$. Cadmium stearate can be regenerated according to reaction (10.33). This is desirable because $BaCl_2$ is a less active catalyst for dehydrochlorination than is $CdCl_2$. A significant synergistic effect can be obtained by using some mixed stabilizers. The stabilizing efficiency of the 0.5/0.5 mixture is higher than that of the separately used (1 and 1) components (Figure 10.11).

Calcium/Zinc Stabilizers

Fatty acid salts of calcium and zinc are nontoxic, FDA sanctioned stabilizers; thus they are often used in food contact applications (packaging films, bottles, etc). The optimal synergistic ratio depends on the aim of stabilization. In oleates, a Ca:Zn ratio of 2:3 was suggested when prevention of the initial coloration is important, while a 4:1 ratio should be used when a long lifetime is more important.

Synergistic behavior strongly depends on degradation conditions. As shown in Figure 10.9 there is an optimum in the length of the induction period at about Ca:Zn = 7:1 when the degradation is carried out in solution. However, as shown in Figure 10.10, the stability is far less with a 7:1 mixture than with calcium stearate alone when the degradation is carried out under dynamic conditions.

Mixed Ca/Zn stabilizers are used as powders, pastes, or liquids. The latter are widely used in plastisol compounds. A typical Ca/Zn-stabilized PVC compound for food packaging soft films consists of 100 parts PVC, 30–70 parts plasticizer,

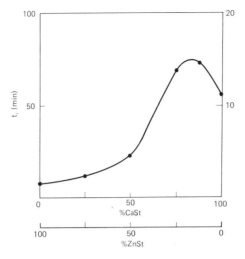

Figure 10.9. Length of induction period (t_i) of PVC degradation in solution as a function of the molar composition of Ca/Zn stearate stabilizer (2% PVC in 1,2,4-trichlorobenzene, 200°C, under argon). Total amount of additive corresponds to a theoretical HCl accepting capacity of 2 mmol HCl/mol monomer unit.

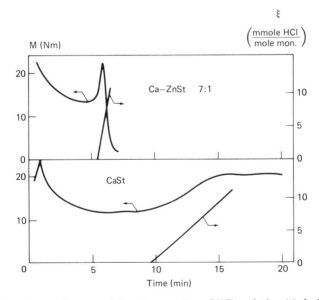

Figure 10.10. Change of torque (M) and conversion of HCl evolution (ξ) during dynamic PVC degradation (65 g PVC in HAAKE Rheomix 600 internal mixer, 190°C set temperature, 50 rpm). Total amount of stabilizer corresponds to a theoretical HCl binding capacity of 5 mmol HCl/mol monomer unit.

2-3 parts Ca/Zn stabilizer, 1-2 parts epoxidized soybean oil, and 0.1-0.2 part stearic acid.

The efficiency of Ca/Zn stabilizers can be significantly enhanced by using auxiliary *phosphite chelators* in combination with them. Excellent chelators are the various 4,4'-isopropylidenediphenol alkyl phosphites with alkyl groups of $C_{12}-C_{15}$. These phosphites also increase the water resistance of the compound. Phosphite chelators can also be used advantageously in combination with Ba/Cd solid stabilizers.

Barium/Cadmium Stabilizers

Powders and liquids of Ba/Cd and Ba/Cd/Zn fatty acid salts are the most widely used PVC heat stabilizers. The application of zinc in Ba/Cd stabilizer packages inhibits initial color formation. The package — with or without Zn — usually also includes some kind of *epoxy plasticizer*. Epoxidized linseed and soybean oils are FDA sanctioned plasticizers. These and other epoxy compounds (octyl epoxy stearates and oleates) not only function as plasticizers but also improve the long-term heat stabilizing efficiency of the metal salt additive.

Synergism is also characteristic of Ba/Cd stabilizers. As shown in Figure 10.11, polyene formation (color development) is faster in the presence of

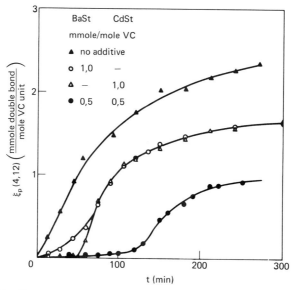

Figure 10.11. Change of the amount of double bonds in 4-12 long polyenes, $\xi_p(4,12)$, as a function of degradation time, for unstabilized and Ba/Cd stearate stabilized PVC, in solution (2% PVC in 1,2,4-trichlorobenzene, 200°C, under argon).

barium stearate than with the cadmium salt. With the 0.5/0.5 mixture, the induction period is twice as long as when the cadmium stearate is applied alone.

Liquid Ba/Cd (and Ba/Cd/Zn) caprylates, phenolates, benzoates, and naphthenates provide clear products, free from plateout and compatibility problems, and are less expensive than solid salts. The solid stearates, laurates, and palmitates provide higher heat stability. These powdery stabilizers find wide application in semirigid and rigid compounds.

A typical PVC compound for seat covers consist of 100 parts PVC, 25-60 parts plasticizer, 1.5-3 parts liquid Ba/Cd stabilizer, 2-3 parts epoxidized soybean oil, 0.2-0.4 part internal and 1 part external lubricant, 0.2-0.4 part UV absorber, and 0-2 parts pigment.

Lead Stabilizers

Electrical insulation and jacketing for wires and cables are generally made from PVC compounds stabilized by lead stabilizers. These mainly consist of basic salts such as dibasic lead carbonate ($2PbO \cdot PbCO_3$), dibasic lead phosphite ($2PbO \cdot PbHPO_3$), and tribasic lead sulfate ($3PbO \cdot PbSO_4$). Organic lead salts such as neutral and dibasic lead stearate, dibasic lead phthalate, lead maleate, as well as their combinations with inorganic salts, are also widely used. Figure 10.12 shows how tribasic lead sulfate (tribase) and lead stearate influence PVC thermal dehydrochlorination. The presence of equivalent amounts of stearic acid increases the stabilization efficiency of the tribase to about the level

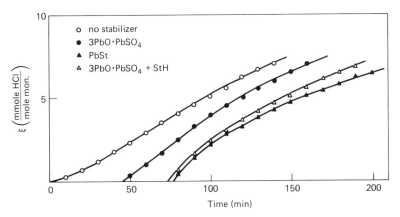

Figure 10.12. Dehydrochlorination conversion (ξ) as a function of degradation time for unstabilized and lead stabilized PVC in solution (2% PVC in 1,2,4-trichlorobenzene, 200°C, under argon). Stabilizers: tribasic lead sulfate (tribase, $3PbO \cdot PbSO_4$), lead stearate (PbSt), and tribase + stearic acid. The amount of additive is equivalent to a theoretical HCl accepting capacity of 2 mmol HCl/mol monomer unit.

of lead stearate. However, stearic acid is usually present as a lubricant in PVC compounds stabilized by inorganic lead salts. Thus, the surprisingly high stabilization efficiency observed in practical application of these additives can probably be explained by the formation of lead stearate from the salt and acid. The transformed stabilizer can function not only as an HCl acceptor but also according to the Frye-Horst mechanism.

Examples of typical lead-stabilized PVC compounds are (a) for insulation masses (up to 1 kV): 100 parts PVC, 30–60 parts plasticizer, 2–4 parts tribase, 0.5–1.5 parts neutral lead stearate, and 20–40 parts chalk; and (b) for pressure tubes: 100 parts PVC, 0.5–1.0 part tribase, 0.5–1 part dibasic lead stearate, 0.2–0.4 part calcium stearate, and 0.2–0.5 part stearic acid.

Organotin Compounds

The most efficient PVC heat stabilizers are the dialkyltin mercaptides and the dialkyltin salts of carboxylic acids. These compounds replace labile chlorines in the polymer with more stable groups (Frye-Horst mechanism). In addition, organotin mercaptides are able to deactivate hydroperoxide groups (hydroperoxide decomposer mechanism). These compounds can also function as HCl acceptors and in some cases (when containing maleic acid derivatives) may exert dienophilic activity as well. Nonradical hydroperoxide decomposing ability is important in thermo-oxidative and photooxidative PVC degradation. Diels-Alder addition of the dienophilic group onto the dienes (or diene parts of the polyenes) can decrease polyene length and discoloration.

Dialkyltin mercaptides, including thioglycolates such as dibutyltin diisooctylthioglycolate,

$$(Bu)_2 Sn(S-CH_2-\overset{\overset{\textstyle O}{\textstyle \|}}{C}-O-isooctyl)_2 \qquad (10.34)$$

are especially active because they exchange allylic chlorine very rapidly. It is assumed that the exchange reaction can even compete with the allyl-activated steps of zip dehydrochlorination. Dialkyltin mercaptides can quite generally be applied to PVC's of any origin and insure good initial color, long-lasting stability, and high transparency. Unfortunately, tin mercaptide stabilizers are fairly expensive and usually rather malodorous. Odor can be eliminated by using alkyl groups containing branching in the γ-position to the S atom, e.g., $-S-CH_2-CH_2-CH(CH_3)_2$. These compounds are usually nontoxic. An FDA sanctioned tin mercaptide is di-*n*-octyltin S,S'-bis(isooctyl mercaptoacetate). A representative compound for food packaging high impact PVC bottles consists of 100 parts PVC, 1–1.5 parts dioctyltin mercaptide, 0.5–1 part internal and 0.5-1 part external lubricants, and 5-10 parts high impact additives (ABS, nitrile rubber, etc.).

Other dialkyltin stabilizers include the salts of maleic acid (and derivatives), lauric acid, and stearic acid. These do not contain sulfur; thus they do not exert antioxidant activity. Nevertheless they are very efficient, generally applicable PVC heat stabilizers, having the advantage over the mercaptides of being odorless and more stable to UV irradiation. Important new organotin stabilizers are the "estertins" of the composition $(R-O-CO-CH_2-CH_2)_2 SnY_2$, and various methyltin derivatives. A typical compound for PVC window frames is as follows: 100 parts PVC, 2-2.5 parts organotin stabilizer, 0.1 part antioxidant, 0.2-0.4 part UV stabilizer, 0.6-1.2 parts internal and 0.6-1.2 parts external lubricants (paraffin wax and oxidized paraffin wax, respectively), 2.4 parts titanium dioxide (rutile; the anatase modification enhances UV decomposition), and 6-12 parts high impact additives.

It should be mentioned that because of the increasingly high price of tin compounds, antimony-based stabilizers, mainly mercaptides, have recently been developed. Their main field of application is in the production of rigid PVC pipes. Organoantimony compounds have poor photostability; thus their application is rather restricted.

10.5 OTHER APPLICATIONS

Stabilizers, or rather additives exerting preventive or protective effects against various types of deterioration, are widely used in polymers. Some of them will be discussed in the following sections. Other additives, such as those used for protection against high energy radiation, and as antifatigue agents, etc., are usually stabilizers which act, on the whole, as antioxidants, radical scavengers, etc., thus, they will not be treated separately.

Stabilization Against Burning — Flame Retardants

Fire retardant additives can act in various ways in a polymer. In the precombustion pyrolysis, they can influence the type and the amount of volatiles. For example, *molybdenum trioxide* reduces the benzene yield during PVC degradation, thus exerting a smoke suppression effect. In the burning material, these additives can retard gas-phase combustion processes by terminating free radicals or by functioning as diluents. Decomposing *halogenated flame retardants* (chlorine and bromine compounds such as chloroparaffins or hexachlorocyclopentadiene derivatives) form free radicals which "poison" the flame by terminating the free radicals in the flame. Another mechanism of flame retardation occurs when the additive eliminates noncombustible gases which can hinder the access of oxygen to the flame. Compounds such as ammonium sulfamate are used in paper but are rarely used in synthetic polymers.

Phosphorus and boron compounds act in the solid phase. By melting they form a coating on the polymer which renders material and heat transfer difficult. In some cases, the flame retardant functions as a solvent for the polymer, lowering its melting temperature: the molten polymer can flow away without inflammation. *Antimony trioxide* is an effective fire retardant for PVC, but it is applied, often in combination with halogenated additives, in other polymers as well. In the presence of halogenated compounds (e.g., PVC), gaseous antimony chlorides and oxychlorides are formed which dilute the combustible gases and can reduce the flame spread. When heat generation is high, they probably act by free radical termination.

Antiozonants

Ozone degradation is especially important in unsaturated polymers. The protection of products against deterioration by ozone (static or dynamic ozone cracking) is an old problem in the rubber industry.

Although some excellent antioxidants are only poor antiozonants and vice versa, the compounds which show the best antiozonant activity belong to the family of conventional chain breaking antioxidants. Thus, *p-phenylenediamine derivatives* are widely applied as antiozonants. Other amines, phenols, and sulfur compounds also find antiozonant application. The mechanism of stabilization is not completely understood. It is important that ozone reactions occur only on the material surface. The additive may scavenge ozone at the surface before it can react with the polymer. Antiozonants may also react with the ozonides and other ozonolysis products. It is assumed that cracking is a result of the physical changes caused by transformation of the ozonolysis products; thus the reaction of the antiozonant with these products may prevent deterioration.

Various *waxes* are used to inhibit ozone deterioration. Waxes migrate to the surface and provide it with a less susceptible coating. The combination of waxes and N,N'-disubstituted *p*-phenylenediamines provides very efficient ozone protection. The wax assists in the diffusion of the antiozonant to the surface.

Stabilization Against Biodegradation

In some polymers, additives are used which may prevent biodegradation caused by fungi. Some "fungicides" are copper 8-quinolinolate/toluenesulfonamide condensates in PVC, 2,2'-thiobis(4,6-dichlorophenol) in polyethylene, and 8-hydroxyquinoline in polyurethanes. Some organotin compounds have been reported as antifungal additives. Organomercury compounds (e.g., phenylmercuric salicylate, dissolved in tricresyl phosphate) may be used not only as fungicides but also as bactericides. Their use is restricted because of their toxicity.

REFERENCES

10.1. Allen, N. S. and McKellar, J. F. Photostabilization of commercial polypropylene by a hindered amine stabilizer. *J. Applied Polymer Sci.* **22**:3277 (1978).

10.2. Allen, N. S., McKellar, J. F. and Wilson, D. Photostabilization of commercial polypropylene by piperidine compounds: the role of stable free radicals. *Polymer Degradation and Stability* **1**:205 (1979).

10.3. Allen, N. S. Photostabilizing performance of a hindered piperidine compound in polypropylene film: anti-oxidant–light stabilizer effects. *Polymer Degradation and Stability* **2**:129 (1980).

10.4. Ayrey, G., Head, B. C. and Poller, R. C. The thermal dehydrochlorination and stabilization of poly(vinyl chloride). *J. Polymer Sci.: Macromolecular Reviews* **8**:1 (1974).

10.5. Bálint, G., Kelen, T., Rehák, Á. and Tüdös, F. Some investigations on the photostabilization of polypropylene. *Reaction Kinetics and Catalysis Letters* **4**:467 (1976).

10.6. Bálint, G., Kelen, T. Tüdös, F. and Rehák, Á. Effect of hindered piperidine derivatives on polypropylene photooxidation. *Polymer Bulletin* **1**:647 (1979).

10.7. Bálint, G., Rockenbauer, A., Kelen, T., Tüdös, F. and Jókay, L. ESR study of polypropylene photostabilization by sterically hindered piperidine derivatives. *Polymer Photochemistry* **1**:139 (1981).

10.8. Carlsson, D. J. and Wiles, D. M. Photostabilization of polypropylene. II. Stabilizers and hydroperoxides. *J. Polymer Sci.–Chem.* **12**:2217 (1974).

10.9. Carlsson, D. J., Garton, A. and Wiles, D. M. The photostabilization of polyolefines. In (Scott, G., ed.) *Developments in Polymer Stabilization–1.* London: Applied Science Publishers, 1979.

10.10. Frye, A. H. and Horst, R. W. The mechanism of poly(vinyl chloride) stabilization by barium, cadmium and zinc carboxylates. I. Infrared studies. *J. Polymer Sci.* **40**:419 (1959).

10.11. Gugumus, F. Developments in the u.v.-stabilization of polymers. In (Scott, G., ed.) *Developments in Polymer Stabilization–1.* London: Applied Science Publishers, 1979.

10.12. Gupta, S. N., Kennedy, J. P., Nagy, T. T., Tüdös, F. and Kelen, T. Heat degradation of PVC stabilized by treatment with alkylaluminum compounds. *J. Macromolecular Sci.–Chem.* **A12**:1407 (1978).

10.13. Iván, B., Kennedy, J. P., Kelen, T. and Tüdös, F. Cyclopentadienylation of PVC: characterization and thermal and thermo-oxidative degradation studies. *J. Polymer Sci.–Chem.* **19**:9 (1981).

10.14. Michel, A. Mechanism of poly(vinyl chloride) stabilization by studies with model compounds. *J. Macromolecular Sci.–Chem.* **A12**:361 (1978).

10.15. Nagy, T. T., Kelen, T., Turcsányi, B. and Tüdös, F. Degradation of barium-cadmium stabilized PVC in solution. *Polymer Bulletin* **2**:749 (1980).

10.16. Nass, L. I. Actions and characteristics of stabilizers. In (Nass, L. I., ed.) *Encyclopedia of PVC*, Vol. 1, Ch. 9. New York: Marcel Dekker, 1976.

10.17. Poller, R. C. *The Chemistry of Organotin Compounds.* New York: Academic Press, 1970.

10.18. Pospíšil, J. Chain-breaking anti-oxidants in polymer stabilization. In (Scott, G., ed.) *Developments in Polymer Stabilization–1*. London: Applied Science Publishers, 1979.

10.19. Scott, G. Mechanism of polymer stabilization. *Pure and Applied chemistry* **30**:267 (1972).

10.20. Shelton, J. R. and Vincent, D. V. Retarded autoxidation and the chain-stopping action of inhibitors. *Journal of the American Chemical Society* **85**:2433 (1963).

10.21. Starnes, W. H., Jr. and Plitz, I. M. Chemical stabilization of poly(vinyl chloride) by prior reaction with di(*n*-butyl)tin bis(*n*-dodecyl mercaptide). *Macromolecules* **9**:633 (1976).

Index

Abrasion, 158
ABS (*see* Acrylonitrile-butadiene-styrene copolymer)
Absorption coefficient, 39, 132, 138, 148
Acetaldehyde, 56, 134
Acetate group elimination, 74
Acetone, 134, 146
Acidimetric analysis, 26
Acoustic energy, 161
Acrylonitrile-butadiene-styrene copolymer, 186, 189
Acrylonitrile derivatives, 188
Actinic radiation, 137, 185
Actinometers, 34
Activation energy, 17, 19, 46, 74, 143, 185
Active hydrogen, 45, 182
Additives, 7, 115, 155, 168
Agricultural applications, 148, 152
Alcohol formation, 49, 77, 113
Alcoholysis, 59, 77
Aldehydes
  formation of, 113, 143
  IR of, 24
Aldose group, 59
Alkoxy radicals, 77, 111, 112
Alkylaluminium compounds, 193
Alkyl radicals, 108, 111, 117, 118
Allyl activation, 75, 80–83, 95, 114, 193
Aluminium layer, 186
AM-104, 188
AM-300, 188
Amines, 148, 177, 181–183, 190, 191, 200
Ammonium sulfamate, 200

Amorphous region, 114
Anatase, 199
Andrews' theory, 165
Anisothermal methods, 14–21, 115
Anthracene, 148, 175
Anthraquinone, 148
Antifatigue agents, 199
Antimony-based stabilizers, 199
Antimony chloride, 200
Antimony oxychloride, 200
Antimony trioxide, 200
Antioxidants, 175–185, 191, 198, 200
Antiozonants, 200
Arrestive stabilization, 174
Arrhenius equation, 17, 40, 56
Aspergillus niger, 153
ASTM D 1435 (weathering), 11
ASTM D 1924 (biodegradation), 153
Autocatalysis, 92, 107
Autoxidation
  of polyenes, 98
  of polyolefins, 107–136
  of PVC, 96–102
Azobisisobutyronitrile, 97
Azo compounds, 97, 148

Bacteria, 152
Bactericides, 200
Ball mill apparatus, 158
Banbury mixers, 161
Baramboim's theory, 167
Barium/cadmium stabilizers, 184, 196

Barium/cadmium/zinc stabilizers, 196
Barium stearate, 194, 196
BAS (*see* Basic autoxidation scheme)
Basic autoxidation scheme, 107
Beer-Lambert's law, 137
Benzene formation, 91, 199
Benzophenones, 148, 187
Benzotriazoles, 187
Benzoylferrocene, 149
Benzyl radical, 180
BHT (*see* Butylated hydroxytoluene)
Bifunctional reaction, 112, 134
Bimolecular deactivation, 140
Bimolecular termination, 113, 123, 126, 129
Biodegradable polymers, 152, 155
Biodegradation, 152–156
Biodeterioration, 152
Biological degradation (*see* Biodegradation)
Bioresistant materials, 152
Bishydrazone, 183
Blends, 6
Bolland-Gee scheme of oxidation, 107
Bond dissociation energy, 3, 4–6, 58, 73, 74, 109, 111, 137, 141, 142
Boron compounds, 200
Brabender test, 102
Bueche's theory, 168
Burning, 199
Butylated hydroxytoluene, 177, 185
Butyraldehyde, 134

Cadmium stearate, 194, 196
Cage effect, 109–111
Calcium/zinc stabilizers, 184, 194
Carbon black, 175, 186
Carbon dioxide formation, 153
Carbon fiber, 49
Carbonization, 49
Carbon monoxide formation, 143, 146
Carbonyl groups
  decomposition of, 142
  effect of, 116
  formation of, 132–135, 189–191
  IR of, 24, 132
Carboxyl groups
  formation of, 113, 135
  IR of, 24, 132

Cavitation, 163, 171
Ceiling temperature, 1, 54–57
Cellulose, 59, 154
Chain branching, 107, 112, 184
Chain breaking antioxidants, 112, 175–182, 191, 200
Chain scission, 43–78, 134, 142, 146, 153, 163
Chain transfer, 45, 76
Chalk, 198
Charlesby-Pinner plot, 70, 71, 86, 96, 104
Chelate formation, 183, 187, 196
Chemical bonds (*see* Bond dissociation energy)
Chemical modification, 7, 193
Chemiluminescence, 26
Chemomechanics, 157
Chlorinated ethylene-propylene copolymers, 85
Chlorinated natural rubber, 18
Chlorobutyl rubber, 85
2-Chloro-2-nitroso-propane, 149
Chloroparaffins, 199
Chloroprene rubber, 75
Chromophores, 144, 147–150
[14]C-labeled carbon dioxide, 153
Coatings, 186
Colony growth, 153
Compatibility, 174
Concentration dependence, 126, 128
Conductometry, 37, 78
Conformational flexibility, 155
Conjugated molecules, 175
Contaminants, 7, 115
Controlled-release drugs, 152
Control of service life, 148
Copolymers, 6, 47, 75, 138, 149, 173, 186, 199
Copper 8-quinolinolate/toluenesulfonamide condensates, 200
Cracking, 157
Crack propagation, 165
Crazing, 157
Critical concentration, 185
Critical film thickness, 115
Critical molecular weight, 167
Cross-link density, 86, 163, 165
Cross-linking, 9, 68–71, 85–89, 96, 104
Crushing, 158

Crystallinity, 114, 163
Cultured fungi and bacteria, 153
Curie-point pyrolyzer, 35
Cutting, 158
Cyanox 2246, 178
Cyasorb UV 531 (*see* Hydroxybenzo-
 phenones)
Cyclic peroxides, 98
Cyclization of polyenes, 84, 89–91
Cyclopropane formation, 84

Deactivation, 138–140
Defect accumulation, 166
Defects (*see* Weak sites)
Degenerate chain branching, 102, 112, 184
Degradation without chain scission, 73–105
Degree of polymerization, 8, 50, 61, 64,
 69, 104, 114, 146, 170
Dehydrobromination, 75
Dehydrochlorination, 48, 71, 75, 78–85,
 99–102, 192–199
Dehydrocyanation, 48, 75
Depolymerization, 1, 43–57, 76
Depolymerization tendency, 43–50, 58
Depropagation (*see* Propagation of depoly-
 merization)
Deuterated samples, 109, 181, 182
Dialkyltin mercaptides, 193, 198
Dibasic lead carbonate, 197
Dibasic lead phosphite, 197
Dibasic lead phtalate, 197
Dibasic lead stearate, 197
Dibutyltin diisooctylthioglycolate, 198
Dibutyltin dilaurate, 155
Dicumyl peroxide, 128
Diels-Alder reaction, 88, 198
Dienophiles, 88
Differential scanning calorimetry, 15
Differential thermal analysis, 15, 115
Diffusion, 7, 111, 114, 146
Diketones, 148
Dilauryl thiodipropionate, 184
2,5-Dimethylfuran, 190
Di-*n*-octyltin S,S′-bis(isooctyl mercapto-
 acetate), 198
Dioxolane, 6, 47, 173
Diphenylmethyllithium, 193
Diphenylpicrylhydrazyl, 163, 175
Dipole interactions, 163

Disulphides, 149
Dosimetry, 33
Double bonds
 effect of, 114, 150
 formation of, 75
Drilling, 158
DSC (*see* Differential scanning calorimetry)
DTA (*see* Differential thermal analysis)
DTG (*see* Thermogravimetry)
DTMA (*see* Thermomechanical analysis)
Dyes, 149, 187
Dynamic PVC degradation, 102–105

Ecolyte copolymers, 154
Elastic materials, 165
Electron paramagnetic resonance (*see*
 Electron spin resonance)
Electron spin resonance, 30, 158, 162, 192
Elimination of side groups, 73–105
EMMA device, 12
EMMAAQUA device, 12
Energy dissipation, 165
Energy transfer, 138–140, 148, 188
Enol/ketone end group, 143
Entanglements, 163, 168
Enthalpy relaxation, 163
Entropy, 54
Entropy relaxation, 163
Enzymes, 152, 154
Epoxidized linseed oil, 196
Epoxidized soybean oil, 155, 196, 197
Epoxy plasticizers, 196
EPR (*see* Electron spin resonance)
ESR (*see* Electron spin resonance)
Estertins, 199
Ethylene-carbon monoxide copolymer, 149
Ethylene-propylene copolymer, 85
Ethylene-vinyl ketone copolymer, 154
Exchange energy transfer, 140
Excitation, 137, 186
Excited state, 113, 139, 141, 144, 187
Extinction coefficient (*see* Absorption coef-
 ficient)
Extrapolation of service life, 41
Extruders, 161
Exudation, 174

Failure time, 41

Fatigue, 157, 171
FDA (*see* Food and Drug Administration)
Fe(III) acetylacetonate, 149
FE dithiocarbamate, 149
Fe 2-hydroxyacetophenone oxime, 149
Fe(III) 2-hydroxy-4-methylacetophenone, 149
Ferrocene, 149, 188
Film thickness, 115, 137, 145
Fire retardants, 199
Flame poisoning, 199
Flame retardants, 199
Flash pyrolysis, 35
Fluorescence, 27, 139
Flynn's method, 40
Food and Drug Administration sanctions, 174, 194, 196, 198
Formaldehyde, 56
Fracture, 157, 165
Fracture mechanics, 165, 171
Freeman-Carrol method, 17
Freezing and thawing, 161
Frenkel's theory, 167
Friction, 158
Frye-Horst mechanism, 194, 198
Fungi, 152
Fungicides, 200

Gas chromatography, 35
Gas-liquid chromatography, 35
Gas-phase reactions, 74
GC (*see* Gas chromatography)
GC/MS (*see* Mas spectrometry)
Gelation (*see* Gel formation)
Gel formation, 60, 69, 85, 96, 104, 163
Gel permeation chromatography, 37, 85, 163
Gel point, 69, 86, 104
Gibbs free energy, 54
GLC (*see* Gas-liquid chromatography)
Glycine, 155
β-Glycoside bond, 59
GPC (*see* Gel permeation chromatography)
Griffith's theory, 165
Grinding, 158
Grosch, Harwood and Payne's theory, 165
Growth rate, 153

HALS (*see* Hindered amine light stabilizers)
Haze, 174
HCl acceptors, 193, 198
HCl catalysis, 78, 92–96, 105, 193
Head-to-head structure, 62
Health regulations, 174
Heat of polymerization, 46, 54
Heat stabilization of PVC, 184, 193–199
Hexachlorocyclopentadiene derivatives, 199
High energy radiation, 60
High impact modifiers, 169, 199
Hindered amine light stabilizers, 182, 183, 191
Hindered amines, 177, 182, 183, 191
Hindered phenols, 177
Homologous series, 59
Horticultural applications, 148, 152
Howard and Gilroy's method, 41
Humidity, 11, 153
Hydrocarbon oxidation, 107
Hydrogen abstraction, 45, 77, 111, 112, 143, 146, 148, 176
Hydrogen donors, 176
Hydrolytic scission, 59, 153, 154
Hydroperoxide decomposers, 112, 147, 183, 186, 190, 198
Hydroperoxides
    in photooxidation, 141–147, 189–191
    in polyolefin oxidation, 98
    in PVC oxidation, 98
    IR of, 24, 131
    photodecomposition of, 141, 146
    transfer to, 176
Hydroperoxide sequences, 111, 132, 146
Hydroquinone, 148
Hydroxybenzophenones, 187
Hydroxyl groups
    elimination of, 74
    IR of, 24, 131
2-Hydroxyphenylbenzotriazole, 187
8-Hydroxyquinoline, 200

Impact test, 157
Impurities, 115
Induced decomposition, 112
Induction period
    of dehydrochlorination, 78, 79, 195

Induction period (*cont.*)
  of oxidation, 176, 184
  of photooxidation, 189–191
Infrared analysis, 24, 104, 131, 163
Inhibition period, 176, 184
Inhibitors, 175
Initiated oxidation
  of polyenes, 98
  of polyolefins, 109, 128
Initiation
  by antioxidants, 185
  of depolymerization, 43
  of photooxidation, 143
  of polyolefin oxidation, 109, 112
  of PVC degradation, 80
Injection molding machines, 161
Insects, 152
Interfaced Pyrolysis Gas Chromatographic Peak Identification System, 36
Internal conversion (*see* Radiationless transitions)
Internal mixer, 158, 195
Internal stabilization, 173, 193
Internal stress, 7
Intersystem crossing, 139, 187
Intramolecular isomerization, 89
Investigation methods, 10–42
  of biodegradability, 153
  of mechanical degradation, 157–163
Iodometry, 25
Ionol, 178
IPGCS (*see* Interfaced Pyrolysis Gas Chromatographic Peak Identification System)
IR (*see* Infrared analysis)
Irganox 1010, 178
Irganox 1076, 178
Iron chloride, 149
Iron complexes, 149
Irradiation devices, 12, 24, 33, 34
Isobutylene formation, 76
4,4'-Isopropylidenediphenol alkyl phosphites, 196
Isothermal methods, 13

Kausch and Hsiao's treatment, 166
Ketene formation, 77
Ketones
  aromatic, 148

Ketones (*cont.*)
  formation of, 112, 146
  IR of, 24, 132
Kinetic chain length, 9, 53, 80, 111, 146
Kinetics
  of depolymerization, 50–54
  of polyolefin thermal oxidation, 116–131
  of PVC degradation, 80–83, 91, 98, 100
  of random scission, 60–62
Kuhn's treatment of chain scission, 63

Lactones, 24
Lake-Thomas theory, 165
Lead maleate, 197
Lead stabilizers, 197
Life phases, 2
Light absorbers, 147, 182, 186–188, 190, 197
Light absorption, 131
Light scattering, 163
Light screens, 186
Light sources, 32
Light stabilizers, 175, 182
Limiting molecular weight, 167
Lithium aluminium hydride, 193
Living organisms, 152
Loss function, 165
Low angle x-ray scattering, 163
Lubricants, 7, 155, 168, 197, 199

Maleic anhydride, 88
Mark 1413, 188
Mass spectrometry, 35
Mastication, 158, 162, 169
McKee consistometer, 161
Mechanical degradation, 41, 157–172
Mechanochemistry, 157
Mercury lamps, 33
Metal chlorides, 194
Metal-ion deactivators, 182
Metallic impurities, 115, 147
Methane formation, 146
Methanol formation, 49, 77
Methyl acrylate, 46
Methylene blue, 149
Methyl methacrylate, 46, 56

Methyl methacrylate-methyl vinyl ketone copolymer, 149
α-Methylstyrene, 46, 56
Mettler thermoanalyzer, 14
Microorganisms, 148, 152–156
Milling, 158, 170
Mirrors, 12, 34
Mixed stabilizers, 194–197
Mixing chamber, 102, 105
Mobility of radicals, 110, 113, 114
Molecular weight (*see* Degree of polymerization)
Molecular weight distribution, 20, 37, 63–68, 70, 163, 168
Molten state, 113, 123, 161
Molybdenium trioxide, 199
Monomer yield, 8, 45, 60, 146
Montsinger equation, 40
Morphology, 7, 163
Most probable distribution (*see* Random distribution)
MWD (*see* Molecular weight distribution)

Naphtalene, 148
Natural rubber, 18, 154, 162
NBC (*see* Nickel dibutyldithiocarbamate)
N-halides, 149
Nickel dibutyldithiocarbamate, 188, 190
Nitrile group reactions, 48
Nitrile rubber, 199
Nitrogen containing chromophores, 148
Nitroso compounds, 148
Nitroxy radicals, 175, 180, 191
NMR (*see* Nuclear magnetic resonance)
N,N'-disubstituted *p*-phenylendiamines, 200
Nonisothermal kinetic analysis, 17
Nonox CI, 178
Nonradiative transfer, 140
N-phenyl-2-naphtylamine, 182
Norrish reactions, 142
Nuclear magnetic resonance, 32, 163
Number of chain scission, 50, 61, 134, 146
Number of cross-links, 69
Nylon 6, 48, 59
Nylon 7, 48
Nylon tire cords, 181

Octyl epoxy oleates, 196
Octyl epoxy stearates, 196

Odor, 174, 198
Olefins, 190
   formation of, 76, 143
Organoantimony compounds, 199
Organomercury compounds, 200
Organotin compounds, 198, 200
Orientation, 166, 168
Osmometry, 85
Outdoor exposure, 11–13
Oxidation scheme
   basic, 107
   polyolefin, 108, 117, 118
   PVC, 102
Oxygen absorption, 101, 145, 182
   investigation methods of, 21–24
   kinetics of, 116–131
Ozone cracking, 200
Ozone photolysis, 144

Paraffin wax, 199
Particle size, 168, 186
Penetrometer curve, 21
Peroxide decomposers (*see* Hydroperoxide decomposers)
Peroxide initiators, 109, 128
Peroxy radicals, 97, 98, 108, 111, 113, 117, 118, 142, 176
Peterlin model, 163
PGC (*see* Pyrolysis gas chromatography)
Phenolic antioxidants, 175, 177–182, 184, 200
Phenoxy radicals, 175
Phenyl benzoate, 188
p-Phenylenediamine derivatives, 200
Phenylmercuric salicylate, 200
Phenyl salicylate, 187
Phosphines, 112
Phosphites, 112, 183, 196, 197
Phosphorescence, 27, 140
Phosphorus compounds, 200
Photochemical processes, 141–143
Photodecomposition, 141
Photodegradable polymers, 149
Photodegradation, 137–151
Photodissociation, 141
Photolytic scission, 142, 146, 149
Photooxidative degradation, 143–150
   effect of carbonyl groups on, 116
   investigation methods of, 21–35
Photooxidation (*see* Photooxidative degradation)

Photophysical processes, 137–140
Photosensitivity, 138
Photostabilization, 185–192
Phtalocyanine dyes, 187
Pigments, 186, 197
Piperidine-N-oxy radical, 192
Plastic deformation, 165
Plasticizers, 7, 155, 168, 196–198
Plastisol compounds, 194
Plating, 174
Polarimetry, 59
Polluted atmosphere, 3, 144
Pollution of environment, 3, 147
Polyacrylates, 76
Polyacrylonitrile, 48, 75, 147, 154
Polyaldehydes, 47
Polyalkylene oxides, 60
Polyamides, 149, 157
Poly-t-butyl methacrylate, 76
Poly-ϵ-caprolactam (see Nylon 6)
Poly-ϵ-caprolactone, 154, 156
Polycarbonate, 138, 187
Poly-α-chloroacrylonitrile, 49
Polychloroprene, 85
Polycyclic aromatic compounds, 148
Polydioxolane, 59
Polyenes
  distribution of, 82
  formation of, 78–85, 193–199
  oxidation of, 96–102
  reactions of, 85–96
Polyenyl radicals, 97
Polyesters, 138, 154
Polyethylene, 6, 25, 36, 45, 59, 70, 109,
  114, 123, 126, 128, 131–135, 138,
  143, 145, 146, 154, 155, 186, 200
Polyethylene oxide, 6, 60
Polyformaldehyde, 6, 47
Polyisobutylene, 6, 45, 169
Polyisocyanates, 47
Polyisoprene, 6, 148, 149, 181, 182
Polymer-analogous reactions, 73
Polymethacrylates, 49, 76
Polymethacrylonitrile, 49
Polymethyl acrylate, 45, 50
Polymethylene, 59
Polymethyl methacrylate, 6, 43, 45, 49, 76,
  146
Polymethyl α-phenylacrylate, 50
Poly-α-methylstyrene, 6, 45, 146

Polyolefin oxidation, 107–136
Polyoxymethylene (see Polyformaldehyde)
Poly-α-phenylacrylonitrile, 49
Polypropylene, 6, 26, 36, 45, 58, 60, 109,
  111, 114, 115, 131–135, 138, 145, 146,
  154, 185, 187, 189–191, 192
Polypropylene oxide, 6, 60
Polysaccharides, 154
Polystyrene, 6, 44, 45, 62, 111, 114, 115,
  138, 154, 169, 170
Polytetrafluoroethylene, 6
Polyurethanes, 154, 200
Polyvinyl acetate, 77, 154
Polyvinyl alcohol, 77
Polyvinyl chloride, 2–4, 7, 37, 71, 73, 75,
  77–105, 138, 142, 154, 155, 173, 174,
  184, 192, 193–200
Prediction of service life, 40, 148
Pressure dependence
  of oxidation, 123, 132
  of ultrasonic degradation, 169
Preventive antioxidants, 182–185
Preventive stabilization, 173, 182
Primary hydrogen, 111, 114
Primary processes of PVC degradation, 78–
  85
Probability treatment of chain scission, 62–
  68
Processing, 2, 107, 147
Product analysis, 35
Propagation
  of cracks, 165
  of dehydrochlorination, 80–83
  of depolymerization, 45–47
  of oxidation, 97, 111
Propionaldehyde, 134
Propionitrile, 75
Propylene, 56
Propylene-vinyl ketone copolymer, 154
Proteins, 154
Proton exchange, 93
Pseudomonas aeruginosa, 153
Pyrazine, 149
Pyrene, 148
Pyridine, 149
Pyrolysis, 19, 35, 199
Pyrolysis gas chromatography, 35

Quenchers, 186, 188–192
Quenching, 144, 188

Quinone methide, 180
Quinones, 148, 175

Radiation cross-linking, 68, 70
Radiationless transitions, 139, 187
Radiation polymerization, 173
Radiative transfer, 140
Radical conversion, 111
Radical scavengers, 163, 175, 186, 191
Radical stabilization (*see* Radical conversion)
Random cross-linking, 68-71
Random degradation, 48, 58-72, 146
Random distribution, 67
Recovery, 3
Reflection, 186
Reinitiation, 95
Resonance excitation transfer, 140
Resonance stabilization, 44, 114
Rodents, 152
Rubber, 18, 75, 85, 154, 162, 181, 199, 200
Rubbery state, 158
Rutile, 199

Salicylaldehyde, 187
Santonox R, 178
β-Scission, 46, 75, 112, 143
Screens, 186
Scroll masticator, 158
Secondary bonds, 163, 165
Secondary hydrogen, 111, 114
Secondary processes of PVC degradation, 85-96
Segmental interactions, 164, 165
Selenides, 112
Self-destructing plastics, 3, 152
Sensitized photodegradation, 147-150
Service life, 11, 40, 148
Shaking, 161
Shear rate, 167, 171
Side groups, 48, 73-105
Singlet oxygen, 113, 144, 148, 186, 190
Singlet state, 139
Slicing, 158
Slippage, 163, 165, 168
Smoke suppression, 199
Soil burial method, 153

Solar radiation (*see* Sunlight)
Sol phase, 86
Solution degradation, 79, 113, 126, 128, 161
Solvent effect, 169
Stability parameters, 19
Stabilization, 112, 173-202
Stabilizers, 112, 173-202
Starch, 154, 156
Steady state, 52, 81, 100, 118
Stearic acid, 196, 198
Stereo-dependent oxidation, 114
Steric factors, 44
Stirring, 161
Stoichiometric coefficient, 176
Stoichiometric parameter, 80, 84, 117
Strain and stress, 166, 171
Stress corrosion, 7
Stress relaxation, 167
Styrene, 46, 56, 111
Styrene-acrylonitrile copolymer, 138
Styrene-phenyl vinyl ketone copolymer, 149, 154
Submicrocrack generation, 164
Sulfides, 112
Sulfoxides, 112
Sulfur compounds, 200
Summation forms, 66
Sunlight, 11, 12, 137, 138, 185
Surface energy, 165
Surface reactions, 146, 200
Surgical implants, 152
Synergism, 175, 194-197

Tacticity, 7, 114
Taste, 174
TBA (*see* Torsional braid analysis)
Termination
  in burning, 199
  in depolymerization, 47
  in polyolefin oxidation, 113
  in PVC degradation, 80, 83, 84
Tertiary amines, 112
Tertiary hydrogen, 58, 109, 111, 114
Tetrafluoroethylene, 56
Tetrahydrofuran, 56
4-(2,2,6,6-Tetramethylpiperidinyl) sebacate, 191
TG (*see* Thermogravimetry)

Thermal degradation, 6, 48, 49, 59, 62, 77–105
Thermal volatilization analysis, 19
Thermogravimetry, 14, 115
Thermomechanical analysis, 20
Thermo-oxidative degradation
  Investigation methods of, 21–35
  of polyolefins, 107–136
  of PVC, 96–102
2,2′-Thiobis(4,6-dichlorophenol), 200
Thioesters, 184
Thioethers, 112, 184
Thioglycolates, 198
Thiophenols, 180
Thiosulfonates, 112
Tie molecules, 163
Tinuvin 326 (see 2-Hydroxyphenylbenzo-triazole)
Tinuvin 770 (see 4-(2,2,6,6-Tetramethyl-piperidinyl) sebacate)
Titanium dioxide, 149, 199
TMA (see Thermomechanical analysis)
Tobolsky and Eyring's treatment, 165
Topanol OC, 178
Torque measurements, 102, 195
Torque rheometers, 102, 161, 195
Torsional brain analysis, 21
Toughness, 165
Toxicity, 174, 198, 201
Tribase (see Tribasic lead sulfate)
Tribasic lead sulfate, 197
Tri-t-butylphenol, 175
Trichlorosuccinimide, 149
Triethylamine, 190
Trioxane, 56
Triplet state, 139
Trisnonylphenyl phosphite, 183
Tritium-labeled HCl, 93
Turbulent flow, 161
TVA (see Thermal volatilization analysis)

Ultrasonic degradation, 161, 170
Unimolecular termination, 113, 118
Unzipping, 9, 47, 73, 80–83
Urethane linkage, 153

UV absorbers (see Light absorbers)
UV stabilizers (see Light stabilizers)
UV-visible spectrophotometry, 39, 78, 94, 96, 104, 196

Van der Waals' bonds, 163
Vibromill, 158
Vinyl chloride-vinyl acetate copolymer, 138
Vinyl ketone copolymers, 149, 156
Viscometry, 37, 59, 161, 163
Viscosity, 86, 169
Volatile products, 44, 113, 134
Volatilization
  of monomers, 44
  of oxidation products, 113, 134

Waste, 1, 147
Waste disposal, 3, 147, 152
Water formation, 113, 146
Waxes, 199, 200
Weak sites, 2, 7, 62, 75, 80, 85, 114, 164, 174, 193
Wear and tear, 158
Weathering, 11–13
Weatherability, 11, 41
Weather-Ometers, 13
Weight loss, 6, 14, 101, 115
Wood, 186

Xenometer, 34
Xenon lamps, 33
Xenotest device, 13

Yelin equation, 40

Zhurkov and Bueche's treatment, 166
Zhurkov-Bueche equation, 41
Zhurkov model, 164
Zinc dibutyldithiocarbamate, 183, 190
Zinc oxide, 149, 187
Zinc stearate, 155